JAMES BRADY'S

GENERAL CHEMISTRY

PRINCIPLES AND STRUCTURE 5/E

STUDENT SOLUTIONS MANUAL

Larry Peck
Texas A & M University

JOHN WILEY & SONS
New York Chichester Brisbane Toronto Singapore

ISBN 0-471-53316-5
Printed in the United States of America

10 9 8 7 6 5 4 3 2 1

PREFACE

The Student Solutions Manual has been prepared as a supplement to GENERAL CHEMISTRY - PRINCIPLES AND STRUCTURE, FIFTH EDITION by James E. Brady. **This version of the manual contains solutions to all odd-numbered problems found at the ends of the chapters in the book by Brady. Another version of this manual contains solutions to all the problems. Permission of the instructor may be required before the other version can be sold in some locations.**

The textbook by J. Brady is an excellent introduction to modern general chemistry. A thorough understanding of general chemistry should include the ability to work problems and answer questions related to the principles and concepts presented. Many students have difficulty developing the ability to utilize knowledge gained from textual material. Through the use of the textbook and this book, this author hopes that students will improve their problem-solving skills while learning a considerable amount of chemistry. Students are strongly encouraged to attempt each problem and to consult the textbook before referring to the solutions presented in this manual. Development of problem-solving skills requires that the student independently attempt each problem before consulting the text or this solution manual. Knowledge gained and problem-solving skills acquired in a general chemistry course should prove valuable in subsequent science and technical courses as well as assisting to prepare students for a wide variety of professions.

Many problems can be answered (solved) in more than one way. I encourage students to develop solutions that they understand. For consistency I have striven to follow the methods used in the textbook. Where practical, I have tried to include dimensions and to show only the correct number of significant figures; in some problems an extra digit is carried in intermediate steps. Remember, authors can make mistakes too.

I wish to thank Dr. Frank Kolar for his tremendous assistance in proofreading and checking solutions in this book. Special thanks go to my family (Sandra, Molly and Marci) for their typing and support during the production of this book.

Larry Peck
Chemistry Department
Texas A&M university

CONTENTS

1 INTRODUCTION

1.1 A chemical is any of the multitude of things that surround us. "Chemicals are everywhere!" In nature chemicals are all the things one can touch, see or smell. A chemical may be a compound or an element. Mixtures contain more than a single chemical.

1.3 We know that a chemical reaction has occurred between sodium and chlorine when they interact to give a new substance with remarkable changes in properties. Not all chemical reactions are as dramatic as the reaction of sodium and chlorine.

1.5 A law summarizes facts while a theory is a tested explanation of the behavior of nature.

1.7 Qualitative observations lack the numbers associated with quantitative observations. Quantitative observations are more useful because they contain more information.

1.9 (a) mass: **Kilogram, kg** (b) time: **Second, s** (c) temperature: **Kelvin, K**
(d) length: **Meter, m** (e) amount of substance: **Mole, mol**

1.11 N = mass x distance/time^2 = (kilograms x meters)/(seconds)^2
= **kg•m/s^2**

1.13 (a) milligram: **mg** (b) decimeter: **dm** (c) kilosecond: **ks**

(d) microsecond: **μs** (e) centigram: **cg**

1.15 (a) 3.2 **nm** = 3.2 x 10^{-9} m
(b) 42 **mm** = 4.2 x 10^{-2} m
(c) 7.3 **mg** = 0.0073 g
(d) 12.5 **cm** = 125 mm
(e) 3.5 **μL** = 3.5 x 10^{-3} mL
(f) 0.84 **dam** = 840 cm

1.17 In the laboratory we usually use the units of grams and milliliters because they are conveniently-sized units for most laboratory measurements of mass and volume. In some laboratories, milligrams and microliters may better reflect the size of measurements most often used but for most laboratories the units of grams and milliliters are the appropriate units.

1.19 **For the Pferdburper:**

$$\left(\frac{10\text{ km}}{1\text{L}}\right) \times \left(\frac{1\text{ L}}{1.057\text{ qt}}\right) \times \left(\frac{4\text{ qt}}{1\text{ gal}}\right) \times \left(\frac{1000\text{ m}}{\text{km}}\right) \times \left(\frac{39.37\text{ in.}}{1\text{ m}}\right) \times$$

$$\left(\frac{1\text{ ft}}{12\text{ in.}}\right) \times \left(\frac{1\text{ mi}}{5280\text{ ft}}\right) = \mathbf{24\ miles/gal.}$$

Therefore, the Pferdburper is better than the Smokebelcher and its 21 miles/gal.

1.21 $$4.3\text{ miles} \times \left(\frac{5280\text{ ft}}{1\text{ mile}}\right) \times \left(\frac{12\text{ in.}}{1\text{ ft}}\right) \times \left(\frac{1\text{ m}}{39.37\text{ in.}}\right) \times \left(\frac{1\text{ km}}{1000\text{ m}}\right) = \mathbf{6.9\ km}$$

1.23 The freezing point and boiling point of water are chosen as the reference temperatures for the definition of the Fahrenheit and Celsius temperature scales because they are convenient, reproducible constant temperatures.

1.25 (a) 50°F = ?°C $(50°\text{F} - 32) \times \frac{5}{9} = \mathbf{10°C}$

(b) 25°C = ?°F $\left(25°\text{C} \times \frac{9}{5}\right) + 32 = \mathbf{77°F}$

(c) 80 K = ?°C 80 K - 273 = **-193°C**

(d) -40°C = ?°F $\left(-40°\text{C} \times \frac{9}{5}\right) + 32 = \mathbf{-40°F}$

(e) 0 K = ?°F 0 K - 273 = -273°C

$$\left(-273°\text{C} \times \frac{9}{5}\right) + 32 = \mathbf{-459°F}$$

1.27 $\frac{5}{9} \times (6152°\text{F} - 32) = \mathbf{3400°C}$ 3400°C + 273 = **3673 K**

1.29 $\frac{5}{9} \times (98.6°\text{F} - 32.0) = \mathbf{37.0°C}$ $\left(\frac{9}{5} \times 39°\text{C}\right) + 32 = \mathbf{102°F}$

1.31 It is important for scientists to indicate the reliability of measured and calculated quantities so others will know the quality of the experimental data. The reliability is reflected by the number of justifiable significant figures that one reports and usually reflects how the values were obtained.

1.33 (a) 1.0370 g has **5** significant figures.
(b) 0.000417 m has **3** significant figures.
(c) 0.00309 cm has **3** significant figures.
(d) 100.1°C has **4** significant figures.
(e) 9.0010 g has **5** significant figures.

1.35 (a) 1,250 g = $\mathbf{1.25 \times 10^3}$ **g**
(b) 13,000,000 m = $\mathbf{1.3 \times 10^7}$ **m**
(c) 60,230,000,000,000,000,000,000 atoms = $\mathbf{6.023 \times 10^{22}}$ **atoms**
(d) $\mathbf{2.1457 \times 10^5}$ **mg**
(e) $\mathbf{3.147 \times 10}$ **g**

1.37 (a) 3×10^{10} m = **30,000,000,000 m**
(b) 2.54×10^{-5} m = **0.0000254 m**
(c) 122×10^{-2} g = **1.22 g**
(d) 3.4×10^{-7} g = **0.00000034 g**
(e) 0.0325×10^{6} cm = **32,500 cm**

1.39 (a) **14.7 cm** (b) **18.3 cm** (c) **27.5 cm** (d) **33.4 cm** (e) **8.4 cm**

1.41 (a) **1.638×10^{9} m** (b) **2.17×10^{9} m** (c) **4.1970×10^{6} m**
(d) **8.5 x 10 mol** (e) **1.00°C**

1.43 (a) $(12.45 \times 10^{6}\ cm^2) \div (2.24 \times 10^{3}\ cm) =$ **5.56×10^{3} cm**
(b) $822\ m \div 0.028\ hr =$ **2.9×10^{4} m/hr**
(c) $(635.4 \times 10^{-5}\ cm) \div (42.7 \times 10^{-4}\ s) =$ **1.49 cm/s**
(d) $(31.3 \times 10^{-12}\ m) \div (8.3 \times 10^{-6}\ m/s) =$ **3.8×10^{-6} s**
(e) $(0.74 \times 10^{-9}\ mol) \div (825.3 \times 10^{18}\ m^3) =$ **9.0×10^{-31} mol/m^3**

1.45 The equality 5280 ft = 1 mile is an exact relationship and, therefore, one can think of it as having an infinite number of significant figures.

1.47 An **extensive property** is one that depends on the size of the sample used. An **intensive property** is one that is independent of the size of the sample used.

Extensive Properties	Intensive Properties
Force (weight), length, number of atoms, moles	Freezing point, specific gravity, specific heat

1.49 The **mass** of an object is a measure of its resistance to a change in velocity and is a measure of the amount of matter in that object. **Weight** is the force with which an object of a certain mass is attracted by gravity to the earth (or some other body such as the moon). The mass of an object does not vary from place to place; it is the same regardless of where it is measured.

1.51 (a) Physical properties of heptane include: liquid, forms a heterogeneous mixture with water and floats on water because it has a density that is less than that of water.
(b) Chemical properties of heptane include: flammable and vapors burn explosively when mixed with air (oxygen) if a spark is provided.

1.53 Density = 14.3 g/8.46 cm^3 = **1.69 g/cm^3**

1.55 Vol. of cyl. = area x length = πr^2 x length = 3.14(1.24 cm)^2 x 4.75 cm =22.9 cm^3.
Therefore, density = 104.2 g/ 22.9 cm^3 = **4.55 g/cm^3**

1.57 (a) 10.00 g $CHCl_3 \Leftrightarrow$? cm^3

$$10.00 \text{ g x} \left(\frac{1 \text{ mL}}{1.492 \text{ g}}\right) \Leftrightarrow \mathbf{6.702 \text{ mL}}$$

(b) 10.00 mL $CHCl_3 \Leftrightarrow$? g

$$10.00 \text{ mL x} \left(\frac{1.492 \text{ g}}{1 \text{ mL}}\right) \Leftrightarrow \mathbf{14.92 \text{ g}}$$

1.59 (a) **500 numerical values** (b) **105 numerical values**

1.61 Specific gravity of propylene glycol times the density of water = the density of propylene glycol. Density of propylene glycol = 1.04 x 8.34 lb/gal = 8.67 lb/gal. The weight of 10,000 gal of propylene glycol would be 8.67 lb/gal x 10,000 gal
= **8.67 x 10^4 lb**

1.63 **Elements** are the simplest forms of matter that can exist under ordinary chemical conditions. **A compound** consists of two or more elements which are always present in the same proportions. **Mixtures** consist of two or more compounds that do not react chemically and differ from elements and compounds in that they may be of variable composition and do not undergo phase changes at constant temperature.

1.65 **A solution** is a homogeneous mixture and has uniform properties throughout. It consists of one phase.

1.67

Homogeneous	Heterogeneous
sea water	smog
air	smoke
black coffee	club soda (with bubbles)
	ham sandwich

1.69 Since the melting point changed during the melting of the solid, one would conclude that the solid sample was a mixture. The temperature of a pure compound would have stayed constant as it melted. This did not. Therefore, it must be a mixture.

1.71 (a) **Fe** (b) **Na** (c) **K** (d) **P** (e) **Br** (f) **Ca** (g) **N** (h) **Ne** (i) **Mn** (j) **Mg**

1.73 (a) Potassium = **2 atoms**, Sulfur = **1 atom**
(b) Sodium = **2 atoms**, Carbon = **1 atom**, Oxygen = **3 atoms**
(c) Potassium = **4 atoms**, Iron = **1 atom**, Carbon and Nitrogen = **6 atoms each**
(d) Nitrogen = **3 atoms**, Hydrogen = **12 atoms**, Phosphorus = **1 atom**, Oxygen = **4 atoms**
(e) Sodium = **3 atoms**, Silver = **1 atom**, Sulfur = **4 atoms**, Oxygen = **6 atoms**

1.75. Aluminum = **1 atom**, Hydrogen = **24 atoms**, Oxygen = **20 atoms**, Potassium = **1 atom**, and Sulfur = **2 atoms**

1.77 $$(3.6 \times 10^4 \text{ m}) + (5.6 \times 10^7 \text{ cm}) = (3.6 \times 10^4 \text{ m}) + \left[(5.6 \times 10^7 \text{ cm}) \times \left(\frac{10^{-2}\text{ m}}{1 \text{ cm}}\right)\right] =$$

$$(3.6 \times 10^4 \text{ m}) + (5.6 \times 10^5 \text{ cm}) = \mathbf{6.0 \times 10^5 \text{ m}}$$

1.79 4255 tons ⇔ ? ft^3 sea water
Density of sea water = 1.025 x 8.34 lb/gal = 8.55 lb/gal

$$4255 \text{ tons} \times \left(\frac{2000 \text{ lb}}{\text{ton}}\right) \times \left(\frac{1 \text{ gal}}{8.55 \text{ lb}}\right) \times \left(\frac{4 \text{ qt}}{\text{gal}}\right) \times \left(\frac{1 \text{ L}}{1.057 \text{ qt}}\right) \times \left(\frac{1 \text{ ft}^3}{28.32 \text{ L}}\right) \Leftrightarrow$$

$$\mathbf{1.33 \times 10^5 \text{ ft}^3}$$

1.81 Superconducting Material
90 K= ?°C
90 K + (-273) = **-183°C**

-183°C = ?°F
[(-183)(9/5)] + 32 = **-297°F**

Boiling point of liquid nitrogen
77 K or **-196°C** or -321°F

2 ATOMS, MOLECULES, AND MOLES

2.1 1. Matter is composed of tiny indivisible particles called atoms.

2. All atoms of a given element are identical, but differ from the atoms of other elements.

3. A chemical compound is composed of the atoms of its elements in a definite fixed numerical ratio.

4. A chemical reaction merely consists of a reshuffling of atoms from one set of combinations to another. The individual atoms remain intact and do not change.

2.3 Atomic mass

2.5 The currently accepted atomic mass unit is: one atomic mass unit is equal to 1/12th of the mass of one atom of carbon-12.

2.7 **No.** If the compound formed is AB, the relative masses would be one set of values; if AB_2, a different set of values; if A_2B, still a different set of values, etc. To calculate the relative masses, one would need to know the ratio of the atoms in the compound formed.

2.9 $$\frac{16.0 \text{ g oxygen}}{12.0 \text{ g carbon}} \Leftrightarrow \frac{?\text{ u (new assigned mass of oxygen)}}{4.00 \text{ u}}$$

oxygen ⇔ **5.33 u**

$$\frac{32.1 \text{ g sulfur}}{12.0 \text{ g carbon}} \Leftrightarrow \frac{?\text{ u (new assigned atomic mass of sulfur)}}{4.00 \text{ u}}$$

sulfur ⇔ **10.7 u**

2.11 $$24.0 \text{ g H} \times \frac{6.00 \text{ g C}}{1.00 \text{ g H}} \Leftrightarrow \mathbf{1.44 \times 10^2 \text{ g C}}$$

2.13 The law of definite composition can be demonstrated by more than one method. One method is to show that the percentage composition does not change from one sample to the next. Another method is to show a constant ratio of elements. This problem has been solved using the ratio method. Students with more chemical knowledge than has been presented thus far in the textbook may select a third method.

For sample #1: 1.00 g C/6.33 g F = 0.158 g C/1 g F
1.00 C/11.67 g Cl = 0.0857 g C/ 1 g Cl
6.33 g F/11.67 g Cl = 0.542 g F/1 g Cl

For sample #2: 2.00 g C/12.66 g F = 0.158 g C/1 g F
2.00 g C/23.34 g Cl = 0.0857 g C/1 g Cl
12.66 g F/23.34 g Cl = 0.542 g F/1 g Cl

For both samples the ratio of mass of carbon to mass of fluorine is 0.158. The ratio of carbon to chlorine is 0.0857 and the ratio of fluorine to chlorine is 0.542 in both samples. These data support the law of definite composition.

2.15 First, for each sample, calculate the amount of one of the elements per gram of the other element. Then compare the results. Grams of phosphorus per gram of oxygen will be calculated. It would be equally correct to calculate the grams of oxygen per gram of phosphorus and to compare those results.
For sample #1: 0.845 g P/(1.50 g sample - 0.845 g P) = 1.29 g P/1 g of O
For sample #2: 1.09 g P/(2.50 g sample - 1.09 g P) = 0.773 g P/1 g of O
This gives 1.29 g P (#1) to 0.773 g P (#2) or when each is divided by the smallest, they become 1.67 to 1.00; then multiplying by 3 the ratio becomes 5-to-3.

2.17 All three represent a set number of objects; the mole is 6.022×10^{23} things, the dozen is 12 things and the gross means 144 things. There are 6.022×10^{23} things in a mole.

2.19 (a) **2-to-3** (b) **2-to-3**

(c) $2 \text{ mol Al} \times \frac{3 \text{ mol O}}{2 \text{ mol Al}} \Leftrightarrow \mathbf{3\ mol\ O}$

(d) $0.2 \text{ mol Al} \times \frac{3 \text{ mol O}}{2 \text{ mol Al}} \Leftrightarrow \mathbf{0.3\ mol\ O}$

2.21 The atom ratios and the mole ratios have the same values in this question.

(a) **1-to-1**

(b) **1-to-1**

(c) **1-to-3**

(d) **1-to-1**

(e) **1-to-3**

(f) **1-to-3**

2.23 (a) 3

(b) 3

(c) 2

(d) 2

(e) $\frac{3 \text{ mol S}}{1 \text{ mol } Al_2(SO_4)_3}$ or $\frac{1 \text{ mol } Al_2(SO_4)_3}{3 \text{ mol S}}$

(f) $\frac{2 \text{ mol Al}}{1 \text{ mol } Al_2(SO_4)_3}$ or $\frac{1 \text{ mol } Al_2(SO_4)_3}{2 \text{ mol Al}}$

2.25 (a) **SiO_2** (b) **50**

(c) **50** (d) **4.50 moles of Si and 9.00 moles of O**

2.27 $1.50 \text{ mol } Cr_2O_3 \times \frac{3 \text{ mol O}}{1 \text{ mol } Cr_2O_3} = \mathbf{4.50 \text{ mol O}}$

2.29 $1 \text{ mol } Al_2(SO_4)_3 \times \frac{3 \text{ mol } SO_4}{1 \text{ mol } Al_2(SO_4)_3} = \mathbf{3 \text{ mol sulfate}}$

2.31 (a) Mg 1 mol = **24.3 g**

(b) C 1 mol = **12.0 g**

(c) Fe 1 mol = **55.8 g**

(d) Cl 1 mol = **35.5 g**

(e) S 1 mol = **32.1 g**

(f) Sr 1 mol = **87.6 g**

2.33 (a) $50.0 \text{ g Na} \times \frac{1 \text{ mol Na}}{23.0 \text{ g Na}} = \mathbf{2.17 \text{ mol Na}}$

(b) **0.668 mol As**

(c) **0.962 mol Cr**

(d) **1.85 mol Al**

(e) **1.28 mol K**

(f) **0.463 mol Ag**

2.35 (a) $24.3 + 16.0 =$ **40.3**

(b) $40.1 + (2 \times 35.45) =$ **111.0**

(c) $30.97 + (5 \times 35.45) =$ **208.2**

(d) $(2 \times 32.07) + (2 \times 35.45) =$ **135.0**

(e) $(3 \times 22.99) + 30.97 + (4 \times 16.00) =$ **163.9**

2.37 (a) $\frac{1 \text{ mol MgO}}{40.3 \text{ g MgO}}$ or $\frac{40.3 \text{ g MgO}}{1 \text{ mol MgO}}$

(b) $\frac{1 \text{ mol } CaCl_2}{111.0 \text{ g } CaCl_2}$ or $\frac{111.0 \text{ g } CaCl_2}{1 \text{ mol } CaCl_2}$

(c) $\frac{1 \text{ mol } PCl_5}{208.2 \text{ g } PCl_5}$ or $\frac{208.2 \text{ g } PCl_5}{1 \text{ mol } PCl_5}$

(d) $\frac{1 \text{ mol } S_2Cl_2}{135.0 \text{ g } S_2Cl_2}$ or $\frac{135.0 \text{ g } S_2Cl_2}{1 \text{ mol } S_2Cl_2}$

(e) $\frac{1 \text{ mol } Na_3PO_4}{163.9 \text{ g } Na_3PO_4}$ or $\frac{163.9 \text{ g } Na_3PO_4}{1 \text{ mol } Na_3PO_4}$

2.39 $\frac{194.193 \text{ g caffeine}}{1 \text{ mol caffeine}} \times 1.35 \text{ mol caffeine} =$ **262 g of Caffeine**

2.41 $\frac{303.3 \text{ g lead sulfate}}{1 \text{ mol lead sulfate}} \times 6.30 \text{ mol} = 1910 \text{ g}$ or **1.91×10^3 g lead sulfate**

2.43 $242 \text{ g } NaHCO_3 \times \frac{1 \text{ mol } NaHCO_3}{84.01 \text{ g } NaHCO_3} =$ **2.88 mol $NaHCO_3$**

2.45 $85.3\text{ g } H_2SO_4 \times \frac{1\text{ mol } H_2SO_4}{98.08\text{ g } H_2SO_4} = \mathbf{0.870\text{ mol } H_2SO_4}$

2.47 $125\text{ g KCl} \times \frac{1\text{ mol K}}{74.55\text{ g KCl}} \Leftrightarrow \mathbf{1.68\text{ mol K}}$

2.49 $1.00 \times 10^3\text{ g } SO_2 \times \frac{1\text{ mol } SO_2}{64.06\text{ g } SO_2} \times \frac{1\text{ mol } FeS_2}{2\text{ mol } SO_2} \Leftrightarrow \mathbf{7.81\text{ mol } FeS_2}$

2.51 $3.50 \times 10^{17}\text{ atoms C} \times \frac{4\text{ atoms H}}{2\text{ atoms C}} \Leftrightarrow \mathbf{7.00 \times 10^{17}\text{ atoms H}}$

$$3.50 \times 10^{17}\text{ atoms C} \times \frac{1\text{ mol C}}{6.022 \times 10^{23}\text{ atoms C}} \times \frac{12.01\text{ g C}}{1\text{ mol C}} = \mathbf{6.98 \times 10^{-6}\text{ g C}}$$

$$7.00 \times 10^{17}\text{ atoms H} \times \frac{1\text{ mol H}}{6.022 \times 10^{23}\text{ atoms}} \times \frac{1.008\text{ g H}}{\text{mol H}} = \mathbf{1.17 \times 10^{-6}\text{ g H}}$$

2.53 $4.00 \times 10^{-8}\text{ g of } C_3H_8 \times \frac{1\text{ mol}}{44.10\text{ g}} \times \frac{6.022 \times 10^{23}\text{ molecules}}{\text{mol}} \times$

$$\frac{3\text{ atoms C}}{\text{molecule } C_3H_8} \Leftrightarrow \mathbf{1.64 \times 10^{15}\text{ atoms C}}$$

2.55 $5.00 \times 10^{20}\text{ molecules } S_8 \times \frac{8\text{ atoms S}}{\text{molecule } S_8} \times \frac{2\text{ atoms Cu}}{\text{atom S}} \times$

$$\frac{1\text{ mol Cu}}{6.022 \times 10^{23}\text{ atoms}} \times \frac{63.5\text{ g Cu}}{\text{mol Cu}} \Leftrightarrow \mathbf{0.844\text{ g Cu}}$$

2.57 (a) C_6H_6 **92.26% C, 7.74% H**

(b) C_2H_5OH **52.14% C, 13.13% H, 34.73% O**

(c) $K_2Cr_2O_7$ **26.58% K, 35.35% Cr, 38.07% O**

(d) XeF_4 **63.34% Xe, 36.66% F**

(e) $CaCO_3$ **40.04% Ca, 12.00% C, 47.96% O**

2.59 (a) $15.0\text{ g }Fe_2O_3 \times \frac{1\text{ mol }Fe_2O_3}{159.70\text{ g }Fe_2O_3} \times \frac{2\text{ mol Fe}}{\text{mol }Fe_2O_3} \times \frac{55.85\text{ g Fe}}{\text{mole Fe}} \Leftrightarrow \mathbf{10.5\text{ g Fe}}$

(b) $25.0\text{ g }Al_2(SO_4)_3 \times \frac{1\text{ mol }Al_2(SO_4)_3}{342.15\text{ g }Al_2(SO_4)_3} \times \frac{2\text{ mol Al}}{\text{mol }Al_2(SO_4)_3} \times \frac{26.98\text{ g Al}}{\text{mol Al}}$

$\Leftrightarrow \mathbf{3.94\text{ g Al}}$

(c) $16.0\text{ g }Na_2CO_3 \times \frac{1\text{ mol }Na_2CO_3}{106.0\text{ g }Na_2CO_3} \times \frac{2\text{ mol Na}}{\text{mol }Na_2CO_3} \times \frac{22.99\text{ g Na}}{\text{mol Na}} \Leftrightarrow \mathbf{6.94\text{ g Na}}$

(d) $48.0\text{ g }MgCl_2 \times \frac{1\text{ mol }MgCl_2}{95.21\text{ g }MgCl_2} \times \frac{1\text{ mol Mg}}{\text{mol }MgCl_2} \times \frac{24.30\text{ g Mg}}{\text{mol Mg}} \Leftrightarrow \mathbf{12.3\text{ g Mg}}$

2.61 $12.0\text{ g }NH_3 \times \frac{1\text{ mol }NH_3}{17.03\text{ g }NH_3} \times \frac{3\text{ mol H}}{\text{mol }NH_3} \times \frac{1.008\text{ g H}}{\text{mol H}} \Leftrightarrow \mathbf{2.13\text{ g H}}$

2.63 (a) $9.34\text{ g }CO_2 \times \frac{1\text{ mole }CO_2}{44.01\text{ g }CO_2} \times \frac{1\text{ mol C}}{\text{mol }CO_2} \times \frac{12.01\text{ g C}}{\text{mol C}} \Leftrightarrow \mathbf{2.55\text{ g C}}$

$5.09\text{ g }H_2O \times \frac{1\text{ mol }H_2O}{18.02\text{ g }H_2O} \times \frac{2\text{ mol H}}{\text{mol }H_2O} \times \frac{1.008\text{ g H}}{\text{mol H}} \Leftrightarrow \mathbf{0.569\text{ g H}}$

(b) g sample = g O + g C + g H or 4.25 - 2.55 - 0.569 = **1.13 g O**

(c) (2.55 g C/4.25 g sample) x 100 = **60.0% C, 13.4% H, 26.6% O**

2.65 (a) NH_4SO_4 (b) Fe_2O_3

(c) $AlCl_3$ (d) CH

(e) $C_3H_8O_3$ (f) CH_2O

(g) Hg_2SO_4

2.67 The simplest formula is calculated from experimentally observed or measured data obtained by a chemical analysis of the compound. One dictionary defined empirical as "founded upon experiment or experience."

2.69 $S_?O_?$ S = 1.40 g x (1 mol S/32.066 g S) = 0.0437 mol
O = 2.10 g x (1 mol O/15.999 g O) = 0.131 mol

$S_{0.0437}O_{0.131}$ or $S_{0.0437/0.0437}O_{0.131/0.0437}$
= S_1O_3 or SO_3

2.71 moles P = 7.04 g P x (1 mol P/30.97 g P) = 0.227 mol P
moles S = 5.46 g S x (1 mol S/32.07 g S) = 0.170 mol S
$P_{0.227}S_{0.170} = P_{1.34}S_{1.00}$ = P_4S_3

2.73 (Assume a 100 g sample). 14.5 g C x (1 mol C/12.01 g C) = 1.21 mol C
85.5 g Cl x (1 mol Cl/35.45 g Cl) = 2.41 mol Cl
$C_{1.21/1.21}Cl_{2.41/1.21}$ = CCl_2

2.75 1.31 g S and 4.22 g - 1.31 g or 2.91 g Cl
1.31 g S x (1mol S/32.07 g S) = 0.0408 mol S
2.91 g Cl x (1 mol Cl/35.45 g Cl) = 0.0821 mol Cl
$S_{0.0408/0.0408}Cl_{0.0821/0.0408}$ = SCl_2

2.77 63.2 g C x (1 mol C/12.01 g C) = 5.26 mol C
5.26 g H x (1 mol H/1.008 g H) = 5.22 mol H
31.6 g O x (1 mol O/16.00 g O) = 1.98 mol O
$C_{5.26/1.98}H_{5.22/1.98/}O_{1.98/1.98}$ or $C_{2.66}H_{2.64}O_{1.00}$ times 3 = $C_8H_8O_3$

2.79 (a) $4.072 \text{ g AgCl} \times \frac{1 \text{ mol AgCl}}{143.32 \text{ g AgCl}} \times \frac{1 \text{ mol Cl}}{\text{mol AgCl}} \times \frac{35.45 \text{ g Cl}}{\text{mol C}} = 1.007 \text{ g Cl}$

(continued)

2.79 (continued)

(b) From part (a), grams Cl = 1.007 g
Grams Cr = 1.500 - 1.007 = 0.493 g

(c) 1.007 g Cl x (1 mol Cl/35.45 g Cl) = 0.0284 mol Cl
0.493 g Cr x (1 mol Cr/52.00 g Cr) = 0.00948 mol Cr
$Cr_{0.00948/0.00948}Cl_{0.0284/0.00948}$ = $\mathbf{CrCl_3}$

2.81 From the H_2O and CO_2 produced by the 1.35 g sample, one can calculate the % H and % C. From the NH_3 produced from the 0.735 g sample, one can obtain the % N. From the AgCl the % Cl is known. By difference, one can obtain % O. Once the % composition is known, the problem becomes very much like Problems 2.73, 2.74, 2.76 and 2.77.

(a)

$$\% \text{ C: } 0.138 \text{ g } CO_2 \times \frac{1 \text{ mol } CO_2}{44.01 \text{ g } CO_2} \times \frac{1 \text{ mol C}}{\text{mol } CO_2} \times \frac{12.01 \text{ g C}}{\text{mol C}} = 0.0377 \text{ g C}$$

$$\frac{0.377 \text{ g C}}{0.150 \text{ g sample}} \times 100 = \mathbf{25.1\% \ C}$$

$$\% \text{ H: } 0.0566 \text{ g } H_2O \times \frac{1 \text{ mol } H_2O}{18.02 \text{ g } H_2O} \times \frac{2 \text{ mol H}}{\text{mol } H_2O} \times \frac{1.01 \text{ g H}}{\text{mol H}} = 0.00634 \text{ g H}$$

$$\frac{0.00634 \text{ g H}}{0.150 \text{ g sample}} \times 100 = \mathbf{4.23\% \ H}$$

$$\% \text{ N: } 0.0238 \text{ g } NH_3 \times \frac{1 \text{ mol } NH_3}{17.03 \text{ g } NH_3} \times \frac{1 \text{ mol N}}{\text{mol } NH_3} \times \frac{14.01 \text{ g N}}{\text{mol N}} = 0.0196 \text{ g N}$$

$$\frac{0.0196 \text{ g N}}{0.200 \text{ g sample}} \times 100 = \mathbf{9.79\% \ N}$$

$$\% \text{ Cl: } 0.251 \text{ g AgCl} \times \frac{1 \text{ mol AgCl}}{143.3 \text{ g AgCl}} \times \frac{1 \text{ mol Cl}}{\text{mol AgCl}} \times \frac{35.45 \text{ g Cl}}{\text{mol Cl}} = 0.0621 \text{ g Cl}$$

$$\frac{0.0621 \text{ g Cl}}{0.125 \text{ g sample}} \times 100 = \mathbf{49.7\% \ Cl}$$

% O: = 100.0 - 25.1 - 4.23 - 9.79 - 49.7 = **11.2% O**

(b) Empirical Formula = $C_{25.1/12.01}H_{4.23/1.01}N_{9.79/14.01}Cl_{49.7/35.45}O_{11.2/16.0}$
= $C_{2.09}H_{4.19}N_{0.70}Cl_{1.40}O_{0.70}$ = $\mathbf{C_3H_6NCl_2O}$

2.83

$$\text{mass C: } 0.6871 \text{ g } CO_2 \times \frac{1 \text{ mol } CO_2}{44.01 \text{ g } CO_2} \times \frac{1 \text{ mol C}}{\text{mol } CO_2} \times \frac{12.01 \text{ g C}}{\text{mol C}} = \mathbf{0.1875 \text{ g C}}$$

$$\text{mass H: } 0.1874 \text{ g } H_2O \times \frac{1 \text{ mol } H_2O}{18.02 \text{ g } H_2O} \times \frac{2 \text{ mol H}}{\text{mol } H_2O} \times \frac{1.008 \text{ g H}}{\text{mol H}} = \mathbf{0.02097 \text{ g H}}$$

(continued)

2.83 (continued)
mass O = 0.5000 g total - 0.1875 g C - 0.02097 g H = **0.2915 g O**

Empirical Formula = $C_{0.1875/12.01}H_{0.02097/1.008}O_{0.2915/16.00}$ =
$C_{0.01561}H_{0.02080}O_{0.01822} = C_{1.000}H_{1.332}O_{1.167}$
Multiplied by 6 it becomes $\mathbf{C_6H_8O_7}$

Molecular Formula? Empirical formula mass= 192. MM = empirical formula mass. Therefore, the molecular formula is $\mathbf{C_6H_8O_7}$

2.85 For carbon-12, atomic mass = 12.0000
For the isotope of gold, atomic mass = 14.9977 x 12.0000 or 179.972

2.87 $$65.0 \text{ g } CuSO_4{\bullet}5H_2O \times \frac{1 \text{ mol } CuSO_4{\bullet}5H_2O}{249.69 \text{ g } CuSO_4{\bullet}5H_2O} \times \frac{9 \text{ mol O}}{1 \text{ mol } CuSO_4{\bullet}5H_2O} \times \frac{16.0 \text{ g O}}{\text{mol O}}$$

$$= \mathbf{37.5 \text{ g O}}$$

2.89 1 Cr atom ⇔ ? volume

$$\frac{51.996 \text{ g Cr}}{\text{mol Cr}} \times \frac{1 \text{ mol Cr}}{6.022 \times 10^{23} \text{ atom Cr}} \times 1 \text{ atom Cr} \times \frac{1 \text{ cm}^3}{7.20 \text{ g}} \Leftrightarrow \mathbf{1.20 \times 10^{-23} \text{ cm}^3}$$

2.91 $Ag_xS_y \rightarrow BaSO_4$

$$0.8689 \text{ g } BaSO_4 \times \frac{1 \text{ mol } BaSO_4}{233.39 \text{ g } BaSO_4} \times \frac{1 \text{ mol S}}{1 \text{ mol } BaSO_4} \times \frac{32.07 \text{ g S}}{\text{mol S}} \Leftrightarrow 0.1194 \text{ g S}$$

0.9225 g sample - 0.1194 g S ⇔ 0.8031 g Ag

0.1194 g S x (1 mol S/32.07 g S) = 0.003723 mol S

0.8031 g Ag x (1 mol Ag/107.87 g Ag) = 0.007445 mol Ag

$Ag_{0.007445}S_{0.003723}$ = $\mathbf{Ag_2S}$

2.93 (a) **2.00 moles** (b) **0.720 moles** (c) **6.00 moles** (d) **3.00 moles**

3 CHEMICAL REACTIONS AND THE MOLE CONCEPT

3.1 The law of conservation of mass

3.3 The coefficients are: (a) **2,3,1,6** (b) **2,1,1,2,2** (c) **3,2,1,6** (d) **1,3,1,3,3** (e) **1,1,1,1,2**

3.5 The coefficients are: (a) **1,2,1,1** (b) **1,4,2,1,1** (c) **1,3,1,3** (d) **3,1,1,2** (e) **2,9,4,6,2**

3.7 (a) $2.50 \text{ mol } CaC_2 \times \frac{1 \text{ mol } C_2H_2}{\text{mol } CaC_2} \Leftrightarrow \mathbf{2.50 \text{ mol } C_2H_2}$

(b) $0.500 \text{ mol } CaC_2 \times \frac{1 \text{ mol } C_2H_2}{\text{mol } CaC_2} \times \frac{26.0 \text{ g } C_2H_2}{\text{mol } C_2H_2} \Leftrightarrow \mathbf{13.0 \text{ g } C_2H_2}$

(c) $(3.20 \text{ mol } C_2H_2) \times (2 \text{ mol } H_2O/\text{mol } C_2H_2) \Leftrightarrow \mathbf{6.40 \text{ mol } H_2O}$

(d) $28.0 \text{ g } C_2H_2 \times \frac{1 \text{ mol } C_2H_2}{26.0 \text{ g } C_2H_2} \times \frac{1 \text{ mol } Ca(OH)_2}{\text{mol } C_2H_2} \times \frac{74.1 \text{ g } Ca(OH)_2}{\text{mol } Ca(OH)_2}$

$\Leftrightarrow \mathbf{79.8 \text{ g } Ca(OH)_2}$

3.9 (a) $\mathbf{P_4 + 5O_2 \rightarrow P_4O_{10}}$

(b) $0.500 \text{ mol } O_2 \times \frac{1 \text{ mol } P_4O_{10}}{5 \text{ mol } O_2} \Leftrightarrow \mathbf{0.100 \text{ mol } P_4O_{10}}$

(c) $50.0 \text{ g } P_4O_{10} \times \frac{1 \text{ mol } P_4O_{10}}{283.9 \text{ g } P_4O_{10}} \times \frac{1 \text{ mol } P_4}{\text{mol } P_4O_{10}} \times \frac{123.9 \text{ g } P_4}{\text{mol } P_4} \Leftrightarrow \mathbf{21.8 \text{ g } P_4}$

(d) $25.0 \text{ g } O_2 \times \frac{1 \text{ mol } O_2}{32.0 \text{ g } O_2} \times \frac{1 \text{ mol } P_4}{5 \text{ mol } O_2} \times \frac{123.9 \text{ g } P_4}{\text{mol } P_4} \Leftrightarrow \mathbf{19.4 \text{ g } P_4}$

3.11 (a) $35.0 \text{ mol Fe} \times \frac{3 \text{ mol CO}}{2 \text{ mol Fe}} \Leftrightarrow \mathbf{52.5 \text{ mol CO}}$

(b) $4.50 \text{ mol } CO_2 \times \frac{1 \text{ mol } Fe_2O_3}{3 \text{ mol } CO_2} \Leftrightarrow \mathbf{1.50 \text{ mol } Fe_2O_3}$

(c) $0.570 \text{ mol Fe} \times \frac{1 \text{ mol } Fe_2O_3}{2 \text{ mol Fe}} \times \frac{159.7 \text{ g } Fe_2O_3}{\text{mol } Fe_2O_3} \Leftrightarrow \mathbf{45.5 \text{ g } Fe_2O_3}$

(d) $48.5 \text{ g } Fe_2O_3 \times \frac{1 \text{ mol } Fe_2O_3}{159.7 \text{ g } Fe_2O_3} \times \frac{3 \text{ mol CO}}{\text{mol } Fe_2O_3} \Leftrightarrow \mathbf{0.911 \text{ mol CO}}$

(e) $18.6 \text{ g CO} \times \frac{1 \text{ mol CO}}{28.0 \text{ g CO}} \times \frac{2 \text{ mol Fe}}{3 \text{ mol CO}} \times \frac{55.85 \text{ g Fe}}{\text{mol Fe}} \Leftrightarrow \mathbf{24.7 \text{ g Fe}}$

3.13 $1{,}000 \text{ kg } C_6H_5Cl \times \frac{1 \text{ kmol } C_6H_5Cl}{112.6 \text{ kg } C_6H_5Cl} \times \frac{1 \text{ kmol DDT}}{2 \text{ kmol } C_6H_5Cl} \times \frac{354.5 \text{ kg DDT}}{\text{kmol DDT}}$

$\Leftrightarrow \mathbf{1.574 \times 10^3 \text{ kg DDT}}$

3.15 (a) $\mathbf{(CH_3)_2NNH_2 + 2N_2O_4 \rightarrow 4H_2O + 2CO_2 + 3N_2}$

(b) $$50.0\ \text{kg}\ (CH_3)_2NNH_2 \times \frac{1\ \text{kmol}\ (CH_3)_2NNH_2}{60.10\ \text{kg}\ (CH_3)_2NNH_2}$$

$$\times \frac{2\ \text{kmol}\ N_2O_4}{1\ \text{kmol}\ (CH_3)_2NNH_2} \times \frac{92.02\ \text{kg}\ N_2O_4}{\text{kmol}\ N_2O_4} \Leftrightarrow \mathbf{153\ kg\ N_2O_4}$$

3.17 (a) $$20.0\ \text{g Pb} \times \frac{1\ \text{mol Pb}}{207.2\ \text{g Pb}} \times \frac{1\ \text{mol white lead}}{6\ \text{mol Pb}} \times \frac{775.6\ \text{g white lead}}{\text{mol white lead}}$$

$$\Leftrightarrow \mathbf{12.5\ g\ white\ lead}$$

(b) $$14.0\ \text{g}\ O_2 \times \frac{1\ \text{mol}\ O_2}{32.0\ \text{g}\ O_2} \times \frac{2\ \text{mol}\ CO_2}{3\ \text{mol}\ O_2} \times \frac{44.01\ \text{g}\ CO_2}{\text{mol}\ CO_2} \Leftrightarrow \mathbf{12.8\ g\ CO_2}$$

3.19 $$2.40\ \text{ton-mol}\ CaCl_2 \times \frac{1\ \text{ton-mol Ca}}{1\ \text{ton-mol}\ CaCl_2} \times \frac{40.08\ \text{tons Ca}}{1\ \text{ton-mol Ca}} \Leftrightarrow \mathbf{96.2\ tons\ Ca}$$

3.21 $$650\ \text{lb}\ H_2 \times \frac{1\ \text{lb-mol}\ H_2}{2.02\ \text{lb}\ H_2} \times \frac{2\ \text{lb-mol}\ NH_3}{3\ \text{lb-mol}\ H_2} \times \frac{17.04\ \text{lb}\ NH_3}{\text{lb-mol}\ NH_3}$$

$$\Leftrightarrow \mathbf{3.66 \times 10^3\ lb\ NH_3}$$

3.23 (a) $0.40\ \text{mol Fe} \times \frac{2\ \text{mol HCl}}{\text{mol Fe}} \Leftrightarrow$ 0.80 mol HCl required to react with 0.40 mol Fe

Since only 0.75 mol HCl is available, HCl is the limiting reactant.

(b) $0.75\ \text{mol HCl} \times \frac{1\ \text{mol}\ H_2}{2\ \text{mol HCl}} \Leftrightarrow \mathbf{0.38\ mol\ H_2}$

(continued)

3.23 (continued)

(c) Moles of Fe required to react with 0.75 mole of HCl is:

$$0.75 \text{ mol HCl} \times \frac{1 \text{ mol Fe}}{2 \text{ mol HCl}} \Leftrightarrow 0.38 \text{ mol Fe}$$

There is 0.40 mol of Fe present. Therefore, 0.40 - 0.38 or **0.02 mol of Fe will remain after all the HCl has reacted.**

3.25 (a) Limiting reactant? Moles Al? $20.0 \text{ g Al} \times \frac{1 \text{ mol}}{26.98 \text{ g}} = 0.741 \text{ mol Al}$

Moles H_2SO_4? $115 \text{ g } H_2SO_4 \times \frac{1 \text{ mol}}{98.08 \text{ g}} = 1.17 \text{ mol } H_2SO_4$

Select either reactant as the limiting reactant and determine if a sufficient quantity of the other is present. Let's select the H_2SO_4. How many moles of Al are needed to react with it?

$$1.17 \text{ mol } H_2SO_4 \times \frac{2 \text{ mol Al}}{3 \text{ mol } H_2SO_4} \Leftrightarrow 0.780 \text{ mole Al}$$

The Al present is less than that required to react with all the H_2SO_4. **Therefore, Al is the limiting reactant.**

(b) $0.741 \text{ mol Al} \times \frac{3 \text{ mol } H_2}{2 \text{ mol Al}} \Leftrightarrow \mathbf{1.11 \text{ mol } H_2}$

(c) $0.741 \text{ mol Al} \times \frac{1 \text{ mol } Al_2(SO_4)_3}{2 \text{ mol Al}} \times \frac{342.1 \text{ g } Al_2(SO_4)_3}{\text{mol } Al_2(SO_4)_3}$

$$\Leftrightarrow \mathbf{127 \text{ g } Al_2(SO_4)_3}$$

(d) $0.741 \text{ mol Al} \times \frac{3 \text{ mol } H_2SO_4}{2 \text{ mol Al}} \times \frac{98.08 \text{ g } H_2SO_4}{\text{mol } H_2SO_4} \Leftrightarrow 109 \text{ g } H_2SO_4 \text{ needed}$

Excess = 115 g - 109 g = **6 g H_2SO_4**

3.27 First find the limiting reactant. $150 \text{ g } CCl_4 \times \frac{1 \text{ mol } CCl_4}{153.8 \text{ g } CCl_4} = 0.975 \text{ mol } CCl_4$

$$100 \text{ g } SbF_3 \times \frac{1 \text{ mol } SbF_3}{178.8 \text{ g } SbF_3} = 0.559 \text{ mol } SbF_3$$

The needed mole ratio is 3 mol CCl_4 to 2 mol SbF_3. Is 0.975 mol of CCl_4 to 0.559 mol of SbF_3 larger or smaller than the needed 3:2 ratio?

$\frac{0.975}{0.559} = 1.74$ Larger! Therefore, the **SbF_3 is the limiting reactant.**

(a) $0.559 \text{ mol } SbF_3 \times \frac{3 \text{ mol } CCl_2F_2}{2 \text{ mol } SbF_3} \times \frac{120.9 \text{ g } CCl_2F_2}{\text{mol } CCl_2F_2} \Leftrightarrow$ **$101 \text{ g } CCl_2F_2$**

(b) Moles of CCl_4 needed to react with 0.559 mole of SbF_3 is:

$$0.559 \text{ mol } SbF_3 \times \frac{3 \text{ mol } CCl_4}{2 \text{ mol } SbF_3} \Leftrightarrow 0.838 \text{ mol } CCl_4$$

$(0.975 \text{ mol } CCl_4 - 0.838 \text{ mol } CCl_4) \times \frac{153.8 \text{ g } CCl_4}{\text{mol } CCl_4} =$ **$21.1 \text{ g } CCl_4$ excess**

3.29 (a) $0.430 \text{ mol } COCl_2 \times \frac{2 \text{ mol HCl}}{1 \text{ mol } COCl_2} \Leftrightarrow$ **0.860 mol HCl**

(b) $11.0 \text{ g } CO_2 \times \frac{1 \text{ mol } CO_2}{44.01 \text{ g } CO_2} \times \frac{2 \text{ mol HCl}}{\text{mol } CO_2} \times \frac{36.45 \text{ g HCl}}{\text{mol HCl}} \Leftrightarrow$ **18.2 g HCl**

(c) Needed mole ratio is 1:1. Therefore, the $COCl_2$ is the limiting reactant.

$0.200 \text{ mol } COCl_2 \times \frac{2 \text{ mol HCl}}{\text{mol } COCl_2} \Leftrightarrow$ **0.400 mol HCl**

3.31 The theoretical yield is the maximum amount of product that could be produced from a given quantity of reactant if the reaction gave only that product, with no side reactions. The percent yield is a comparison of the yield that is actually obtained to the theoretical yield; it is the efficiency of the reaction. The actual yield is the amount of product that you actually obtain in a given experiment when the reaction is carried out.

3.33 (a) $6.40 \text{ g } CH_3OH \times \frac{1 \text{ mol } CH_3OH}{32.0 \text{ g } CH_3OH} \times \frac{2 \text{ mol } CO_2}{2 \text{ mol } CH_3OH} \times \frac{44.0 \text{ g } CO_2}{\text{mol } CO_2}$

$\Leftrightarrow$ **8.80 g CO_2** = theoretical yield of CO_2

(b) Actual yield given as 6.12 g CO_2

(c) Percentage yield = (6.12/8.80) x 100% = **69.5%**

3.35 (a) Mol CH_3Cl + mol CH_2Cl_2 + mol $CHCl_3$ + mol CCl_4 must equal mol of CH_4 at the start if the total amount of C is to be maintained.

mol CH_4 ? $\frac{20.8 \text{ g } CH_4}{16.04 \text{ g/mol}} = 1.30 \text{ mol } CH_4$

mol CH_3Cl ? $\frac{5.0 \text{ g } CH_3Cl}{50.5 \text{ g/mol}} = 0.099 \text{ mol } CH_3Cl$

mol CH_2Cl_2? $\frac{25.5 \text{ g } CH_2Cl_2}{84.93 \text{ g/mol}} = 0.300 \text{ mol } CH_2Cl_2$

mol $CHCl_3$? $\frac{59.0 \text{ g } CHCl_3}{119.37 \text{ g/mol}} = 0.494 \text{ mol } CHCl_3$

mol CCl_4 ? 1.30 - 0.099 - 0.30 - 0.494 = 0.41 mol CCl_4

mass CCl_4 ? 0.41 mol CCl_4 x 153.8 g/mol = **63 g CCl_4**

(b) If all the available CH_4 had been converted to CCl_4, the theoretical yield would be:

$1.30 \text{ mol } CH_4 \times \frac{1 \text{ mol } CCl_4}{\text{mol } CH_4} \times \frac{153.8 \text{ g } CCl_4}{\text{mol } CCl_4} \Leftrightarrow \mathbf{2.00 \times 10^2 \text{ g } CCl_4}$

(c) Percentage yield = (63/200) x 100% = **32%**

(d) $CH_4 + Cl_2 \rightarrow CH_3Cl + HCl$,

$CH_4 + 2Cl_2 \rightarrow CH_2Cl_2 + 2HCl$

$CH_4 + 3Cl_2 \rightarrow CHCl_3 + 3HCl$,

$CH_4 + 4Cl_2 \rightarrow CCl_4 + 4HCl$

(0.099 mol CH_3Cl x 1) + (0.300 mol CH_2Cl_2 x 2) + (0.494 mol $CHCl_3$ x 3) + (0.41 mol CCl_4 x 4) = 3.82 mol Cl_2

3.82 mol Cl_2 x 70.91 g/mol = **271 g Cl_2**

3.37 Solutions are used to carry out chemical reactions because the reactions will take place much more rapidly when the reactants are in a dissolved state.

3.39 Molar concentration = number of moles of solute/total volume of the solution in liters.

3.41 0.20 M Na_3PO_4 means; $\frac{0.20 \text{ mol } Na_3PO_4}{1.0 \text{ L solution}}$ and/or $\frac{1{,}000 \text{ mL solution}}{0.20 \text{ mol } Na_3PO_4}$

3.43 (a) 1.35 mol NH_4Cl/2.45 L soln. = 0.551 mol/L soln. = **0.551 M NH_4Cl**

(b) 0.422 mol $AgNO_3$/0.742 L soln. = **0.569 M $AgNO_3$**

(c) 3.00×10^{-3} mol KCl/0.0100 L soln. = **0.300 M KCl**

(d) $$\frac{4.80 \times 10^{-2} \text{ g } NaHCO_3}{0.0250 \text{ L}} \times \frac{1 \text{ mol}}{84.01 \text{ g}} = \frac{0.0229 \text{ mol } NaHCO_3}{\text{L soln.}}$$

$$= \mathbf{0.0229 \text{ M } NaHCO_3}$$

3.45 (a) $\frac{0.150 \text{ mol}}{\text{L}} \times 0.250 \text{ L} = \mathbf{0.0375 \text{ mol } Li_2CO_3}$

(b) $$\frac{0.150 \text{ mol } Li_2CO_3}{\text{L}} \times 0.630 \text{ L soln.} \times \frac{73.89 \text{ g } Li_2CO_3}{\text{mol } Li_2CO_3} = \mathbf{6.98 \text{ g } Li_2CO_3}$$

(c) $$\frac{0.0100 \text{ mol } Li_2CO_3}{0.150 \text{ mol } Li_2CO_3/\text{L soln.}} = 0.0667 \text{ L} = \mathbf{66.7 \text{ mL solution}}$$

(d) $$0.0800 \text{ g } Li_2CO_3 \times \frac{1 \text{ mol } Li_2CO_3}{73.89 \text{ g } Li_2CO_3} \times \frac{1 \text{ L}}{0.150 \text{ mol } Li_2CO_3}$$

$$\times \frac{10^3 \text{ mL}}{\text{L}} = \mathbf{7.22 \text{ mL solution}}$$

3.47 $\frac{0.250 \text{ mol}}{\text{L}} \times \frac{2.00 \text{ L}}{1} \times \frac{158.2 \text{ g } Ca(C_2H_3O_2)_2}{\text{mol}} = \mathbf{79.1 \text{ g } Ca(C_2H_3O_2)_2}$

3.49 $\frac{0.150 \text{ mol } MgSO_4}{\text{L}} \times 0.500 \text{ L} \times \frac{1 \text{ mol } MgSO_4{\cdot}7H_2O}{\text{mol } MgSO_4} \times \frac{246.5 \text{ g } MgSO_4{\cdot}7H_2O}{1 \text{ mol } MgSO_4{\cdot}7H_2O}$

$\Leftrightarrow \mathbf{18.5 \text{ g } MgSO_4{\cdot}7H_2O}$

3.51 Add the more dense, concentrated reagent slowly to the water.

3.53 Concentrated NH_3 is 15 **M** (Table 3.1)

15 **M** x V_i = 0.500 **M** x 250 mL

$V_i = \frac{0.500 \text{ M} \times 250 \text{ mL}}{15 \text{ M}}$

V_i = **8.3 mL**

3.55 0.500 **M** x 100 mL = 0.200 **M** x V_f

$V_f = \frac{0.500 \text{ M} \times 100 \text{ mL}}{0.200 \text{ M}} = \mathbf{2.50 \times 10^2 \text{ mL}}$

3.57 Known values: **M** = moles/L or **M** x L = moles

Let ? = L of 1.00 **M** HCl added

$0.600 \text{ M} = \frac{(0.500 \text{ M} \times 0.0500 \text{ L}) + (1.00 \text{ M} \times ? \text{ L})}{(0.0500 \text{ L} + ? \text{ L})}$

0.0300 + 0.600 ? = 0.0250 + ?

0.40 ? = 0.0050

? = 0.0125 L or **12 mL**

3.59 (a) $\frac{0.250 \text{ mol NaBr}}{\text{L}} \times 0.300 \text{ L} = 0.0750 \text{ mol NaBr}$

$(0.400 \text{ mol } AgNO_3/\text{L}) \times 0.200 \text{ L} = 0.0800 \text{ mol } AgNO_3$

Since they react in a 1:1 mole ratio, the **NaBr is the limiting reactant.**

(b) $0.0750 \text{ mol NaBr} \times \frac{1 \text{ mol AgBr}}{\text{mol NaBr}} \times \frac{187.8 \text{ g AgBr}}{\text{mol AgBr}} \Leftrightarrow \mathbf{14.1 \text{ g AgBr}}$

3.61 Let's use the definition of molarity as mmol solute divided by mL solution as a working equation. $\mathbf{M} = \frac{\text{mmol}}{\text{mL}}$ can be used as needed.

(a) $\frac{0.200 \text{ mmol } MgCl_2}{\text{mL}} \times 75.0 \text{ mL} = 15.0 \text{ mmol } MgCl_2$

$15.0 \text{ mmol } MgCl_2 \times \frac{2 \text{ mmol NaOH}}{\text{mmol } MgCl_2} \Leftrightarrow 30.0 \text{ mmol NaOH}$

$0.300 \text{ M} = \frac{30.0 \text{ mmol}}{? \text{ mL}}$ $\quad ? \text{ mL} = 30.0/0.300 = \mathbf{100 \text{ mL}}$

(b) $\frac{0.600 \text{ mmol } MgCl_2}{\text{mL}} \times 50.0 \text{ mL} = 30.0 \text{ mmol } MgCl_2$

$30.0 \text{ mmol } MgCl_2 \times \frac{1 \text{ mmol } Mg(OH)_2}{\text{mmol } MgCl_2} \Leftrightarrow 30.0 \text{ mmol } Mg(OH)_2$

$30.0 \text{ mmol } Mg(OH)_2 \times \frac{1 \text{ mol}}{10^3 \text{ mmol}} \times \frac{58.32 \text{ g } Mg(OH)_2}{\text{mol } Mg(OH)_2} = \mathbf{1.75 \text{ g } Mg(OH)_2}$

(continued)

3.61 (continued)

(c) $\frac{0.200 \text{ mmol MgCl}_2}{\text{mL}} \times 30.0 \text{ mL} = 6.00 \text{ mmol MgCl}_2$

$\frac{0.140 \text{ mmol NaOH}}{\text{mL}} \times 100 \text{ mL} = 14.0 \text{ mmol NaOH}$

The needed mole - or mmol - ratio is 1 of $MgCl_2$ to 2 of NaOH. In this reaction the $MgCl_2$ is the limiting reactant.

$$6.00 \text{ mmol MgCl}_2 \times \frac{1 \text{ mmol Mg(OH)}_2}{\text{mmol MgCl}_2} \times \frac{10^{-3} \text{mol}}{\text{mmol}} \times \frac{58.32 \text{ g Mg(OH)}_2}{\text{mol Mg(OH)}_2}$$

$\Leftrightarrow$ **0.350 g $Mg(OH)_2$**

3.63 $15.0 \text{ mL NaOH} \times \frac{0.750 \text{ mol NaOH}}{\text{L}} \times \frac{1 \text{ L}}{1000 \text{ mL}} = 0.0112 \text{ mol NaOH}$

$0.0112 \text{ mol NaOH} \times \frac{1 \text{ mol H}_2\text{SO}_4}{2 \text{ mol NaOH}} \Leftrightarrow 5.60 \times 10^{-3} \text{mol H}_2\text{SO}_4$

$\frac{5.60 \times 10^{-3} \text{mol H}_2\text{SO}_4}{0.0250 \text{ L}} = 0.224 \text{ M H}_2\text{SO}_4$ used in reaction

$1.40 \text{ M} \times 250 \text{ mL} = 0.224 \text{ M} \times V_f$

$V_f = \frac{1.40 \text{ M} \times 250 \text{ mL}}{0.224 \text{ M}} = \mathbf{1.56 \times 10^3 \text{ mL}}$

3.65 $0.400 \text{ L} \times \frac{0.200 \text{ mol}}{\text{L}} = 0.0800 \text{ mol}$ (solution # 1)

$0.800 \text{ L} \times \frac{0.600 \text{ mol}}{\text{L}} = 0.480 \text{ mol}$ (solution # 2)

(Assume that the volumes are additive)

Final solution is: $\frac{0.0800 \text{ mol} + 0.480 \text{ mol}}{0.400 \text{ L} + 0.800 \text{ L}} = \mathbf{0.467 \text{ M}}$

3.67 (a) If the Mg is the limiting reactant:

$$6.00 \text{ g Mg} \times \frac{1 \text{ mol Mg}}{24.305 \text{ g Mg}} \times \frac{1 \text{ mol H}_2}{\text{mol Mg}} \times \frac{2.02 \text{ g H}_2}{\text{mol H}_2} \Leftrightarrow 0.499 \text{ g H}_2$$

If the HCl is the limiting reactant:

$$0.350 \text{ L} \times \frac{0.800 \text{ mol HCl}}{\text{L}} \times \frac{1 \text{ mol H}_2}{2 \text{ mol HCl}} \times \frac{2.02 \text{ g H}_2}{\text{mol H}_2} = 0.283 \text{ g H}_2$$

Since the amount of HCl will produce less H_2, the HCl solution is the limiting reactant and the answer is the **0.283 g H_2.**

(b) First calculate theoretical yield.

$$0.350 \text{ L} \times \frac{0.800 \text{ mol HCl}}{\text{L}} \times \frac{1 \text{ mol MgCl}_2}{2 \text{ mol HCl}} \times \frac{95.2 \text{ g MgCl}_2}{\text{mol MgCl}_2}$$

$$\Leftrightarrow 13.3 \text{ g MgCl}_2 \text{ (theoretical yield)}$$

Percentage yield = (8.64/13.3 g) x 100% = **65.0%**

3.69 Limiting reactant? HCl? 0.235 L x 0.600 mol/L = 0.141 mol HCl
Na_2CO_3? 0.0940 L x 0.750 mol/L = 0.0705 mol Na_2CO_3

Both are the limiting reactants since they are mixed in a ratio equal to that indicated by the balanced equation. Either could be used to calculate the answer.

$$0.141 \text{ mol HCl} \times \frac{2 \text{ mol NaCl}}{2 \text{ mol HCl}} \times \frac{1}{(.235 \text{ L} + 0.0940 \text{ L})} \Leftrightarrow \frac{0.429 \text{ mol NaCl}}{\text{L soln.}}$$

= **0.429 M NaCl**

4 THE PERIODIC TABLE AND SOME PROPERTIES OF THE ELEMENTS

4.1 The ability to deform when hammered is called malleability. A blacksmith relies on the malleability of iron when forging a horseshoe.

4.3 Three properties of metals, other than malleability and ductility, are: (1) metallic luster, (2) good conductors of heat and (3) good conductors of electricity.

4.5 $2Na + 2H_2O \rightarrow 2NaOH + H_2$

4.7 Since they are good conductors of heat, metals feel hot when left in the sun. As your hand absorbs heat from the metal, heat travels quickly from the neighboring parts of the object to replace the heat your hand absorbed, thus providing more heat for your hand to absorb.

4.9 Tungsten has the highest melting point of any element which accounts for its use as the filament in electrical light bulbs. Mercury has the lowest melting point of any metal. Mercury is the fluid used in some thermometers.

4.11 Graphite and diamond. Both are made up of carbon. Both lack luster, are nonmalleable and nonductile. Graphite is soft and opaque while diamond is transparent and very hard.

4.13 Oxygen and nitrogen

4.15 Metalloids look somewhat like metals but are darker in color. They conduct electricity but not nearly as well as metals. Metalloids are much more like nonmetals than metals.

4.17 Mendeleev left spaces for yet undiscovered elements because there were not known elements with the properties that fit into the pattern.

4.19 Co and Ni, Th and Pa, and U and Np

4.21 One coulomb is equal to the amount of charge that moves past a given point in a wire when an electric current of 1 ampere flows for 1 second.
1 C = 1A x 1s

The charge on a mole of electrons =

$$\frac{1.602 \times 10^{-19}\,\text{C}}{e^-} \times \frac{6.022 \times 10^{23}\,e^-}{\text{mol } e^-} = \frac{\mathbf{9.647 \times 10^{4}\,C}}{\mathbf{mol\ e^-}}$$

4.23 The atomic number is the number of protons in the nucleus of the atom (and e⁻'s in neutral atom).

4.25 $$\frac{+4.8 \times 10^{-19}\,\text{C}}{1.6 \times 10^{-19}\,\text{C/e}^-\ \text{lost}} = +3 \text{ or } \mathbf{Al^{3+}}$$

4.27 Density = m/V and mass = 1.67×10^{-24} g
$V = (4/3)\pi r^3 = (4/3) \times 3.142 \times (0.500 \times 10^{-13}\ \text{cm})^3$
$= (4/3) \times 3.142 \times 1.25 \times 10^{-40}\ \text{cm}^3 = 5.24 \times 10^{-40}\ \text{cm}^3$
Density = 1.67×10^{-24} g/5.24×10^{-40} cm^3 = **3.19×10^{15} g/cm^3**

4.29 Mass of the earth in grams is:

$$6.59 \times 10^{21}\ \text{tons} \times \frac{2000\ \text{lb}}{\text{ton}} \times \frac{454\ \text{g}}{\text{lb}} = 5.98 \times 10^{27}\ \text{g}$$

Use Density = 3.19×10^{15} g/cm^3 from problem 4.27

$$\frac{5.98 \times 10^{27}\ \text{g}}{3.19 \times 10^{15}\ \text{g/cm}^3} = 1.87 \times 10^{12}\ \text{cm}^3$$

Use $V = (4/3)\ \pi\ r^3$ to obtain the radius. $1.87 \times 10^{12} = (4/3)\ \pi\ r^3$
$r = 7.64 \times 10^3$ cm
Diameter would be: $2 \times 7.64 \times 10^3$ cm or 1.53×10^4 cm or **153 meters** or **502 ft** or **0.0951 mile**

4.31 (a) A proton has a mass of 1.007276 u, or approximately 1 u, and a charge of 1+.
(b) A neutron has a mass of 1.008665 u, or approximately 1 u, and a charge of 0.
(c) An electron has a mass of 0.0005486 u, or approximately 0 u, and a charge of 1-

4.33 The mass number is simply the total count of protons plus neutrons and is not quite equal to the atomic mass of an atom which is the atom's actual mass.

4.35 ^{131}Ba has 56 protons, 75 neutrons and 56 electrons.
$^{109}Cd^{2+}$ has 48 protons, 61 neutrons and 46 electrons.
$^{36}Cl^-$ has 17 protons, 19 neutrons and 18 electrons.
^{63}Ni has 28 protons, 35 neutrons and 28 electrons.
^{107}Tm has 69 protons, 101 neutrons and 69 electrons.

4.37 (a) **29** (b) **49** (c) **123** (d) **99** (e) **99**

4.39 ^{10}B (10.01294 u/atom) x 0.196 atom = 1.96 u
^{11}B (11.00931 u/atom) x 0.804 atom = 8.85 u
Average mass of one atom = **10.81 u**
Atomic mass = **10.81 g/mol**

4.41 (34.96885 x ?% ^{35}Cl) +[36.96590 x (100 - ?% ^{35}Cl)] = 35.453 x 100%
(34.96885 x ?% ^{35}Cl) +(3696.590 - 36.96590 x ?% ^{35}Cl) = 3545.3
(36.96590 - 34.96885) ?% ^{35}Cl = 3696.590 - 3545.3
1.99705 ?% ^{35}Cl = 151.3
?% ^{35}Cl = **75.76%** % ^{37}Cl = (100.00 - 75.76)% = **24.24%**

4.43

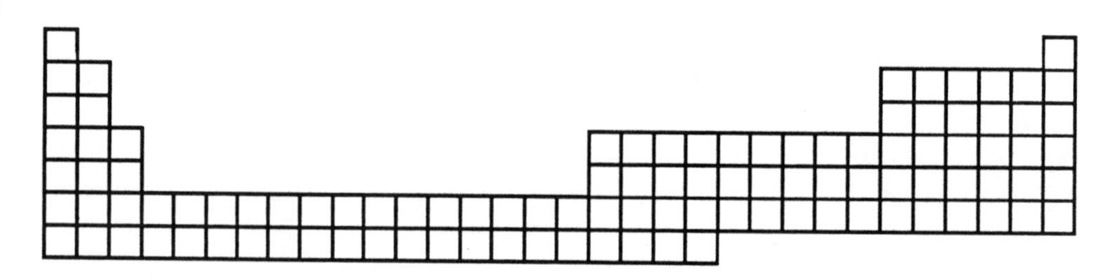

4.45 Mg, Se and Br

4.47 Elements 58 through 71 and 90 through 103

4.49 F_2, Cl_2, Br_2 and I_2

4.51 Ba

4.53 B, Si, Ge, As, Sb, Te, Po and At (Al is not a metalloid)

4.55

	Metal	Nonmetal	Metalloids
Period 4	K, Ca, Sc, Ti V, Cr, Mn, Fe, Co Ni, Cu, Zn, Ga	Se, Br, Kr	Ge, As
Group VA	Bi	N, P	As, Sb

4.57 $RaCl_2$

4.59 A combination reaction is one in which two or more substances combine to form a single product.

4.61 (a) Sr^{2+} (b) Na^+ (c) S^{2-} (d) Al^{3+} (e) Br^-

4.63 (a) Fe^{2+} and Fe^{3+} (b) Cu^+ and Cu^{2+} (c) Sn^{2+} and Sn^{4+}

(d) Zn^{2+} (e) Cr^{2+} and Cr^{3+}

4.65 See Table 4.3 (a) $AuBr$, $AuBr_3$, Au_2O and Au_2O_3
(b) $ZnBr_2$ and ZnO
(c) $CdBr_2$ and CdO
(d) $AgBr$ and Ag_2O
(e) $SnBr_2$, $SnBr_4$, SnO and SnO_2
(f) $PbBr_2$, $PbBr_4$, PbO and PbO_2
(g) $BiBr_3$ and Bi_2O_3

4.67 (a) CN^- (b) ClO_4^- (c) MnO_4^- (d) NO_3^- (e) PO_4^{3-}

(f) OH^- (g) $C_2O_4^{2-}$ (h) $Cr_2O_7^{2-}$ (i) SO_4^{2-} (j) HCO_3^-

(k) SO_3^{2-} (l) NO_2^-

4.69 (a) $CrCO_3$, $Cr_2(CO_3)_3$, $CrCrO_4$, $Cr_2(CrO_4)_3$, $CrSO_3$, $Cr_2(SO_3)_3$, $Cr(C_2H_3O_2)_2$, $Cr(C_2H_3O_2)_3$

(b) $MnCO_3$, $Mn_2(CO_3)_3$, $MnCrO_4$, $Mn_2(CrO_4)_3$, $MnSO_3$, $Mn_2(SO_3)_3$, $Mn(C_2H_3O_2)_2$, $Mn(C_2H_3O_2)_3$

(c) $FeCO_3$, $Fe_2(CO_3)_3$, $FeCrO_4$, $Fe_2(CrO_4)_3$, $FeSO_3$, $Fe_2(SO_3)_3$, $Fe(C_2H_3O_2)_2$, $Fe(C_2H_3O_2)_3$

(d) through (n) Repeat using the other 11 metals.

4.71 Compounds formed between two nonmetals are not held together by an attraction between ions but rather by the sharing of electrons. In BaF_2 the force between positive Ba^{2+} ions and negative F^- ions is the attraction between ions of opposite charge. In H_2 the force is a sharing of electrons.

4.73 PbO and PbO_2

4.75 (a) CF_4 (b) NF_3 (c) OF_2
(d) AsF_3 (e) ClF

4.77 Ionic compounds are more brittle than molecular compounds because the forces in ionic compounds are such that even a small slippage of one part of the solid can cause attractions to change to repulsions. The repulsive forces cause the crystal to break. Molecular solids, on the other hand, are much more flexible since there are no strong attractions or repulsions between molecules.

4.79 (a) CaF_2 (b) $AlCl_3$ (c) TiO_2 (d) NaH
In each case, the ionic compound will have the higher melting point.

4.81 Oxidation number and oxidation state are interchangeable terms.

4.83 The actual rules are given in the textbook.

4.85
(a) K, +1; Cl, +3; O, -2
(b) Ba, +2; Mn, +6; O, -2
(c) Fe, +8/3; O, -2
(d) O, +1; F, -1
(e) I, +5; F, -1
(f) H, +1; O, -2; Cl, +1
(g) Ca, +2; S, +6; O, -2
(h) Cr, +3; S, +6; O, -2
(i) O, 0
(j) Hg, +1; Cl, -1

4.87
(a) **oxidation**
(b) **reduction**
(c) **oxidation**
(d) **oxidation**
(e) **reduction**

4.89

	Oxidized	Reduced	Oxidizing Agent	Reducing Agent
(a)	NaI	$NaIO_3$	$NaIO_3$	NaI
(b)	Cu	HNO_3	HNO_3	Cu
(c)	Cu	HNO_3	HNO_3	Cu
(d)	Cu	H_2SO_4	H_2SO_4	Cu
(e)	SO_2	HNO_3	HNO_3	SO_2

4.91
(a) $Al(NO_3)_3$
(b) $FeSO_4$
(c) $NH_4H_2PO_4$
(d) IF_5
(e) PCl_3
(f) N_2O_4
(g) $KMnO_4$
(h) $Mg(OH)_2$
(i) H_2Se
(j) NaH

4.93
(a) TiO_2
(b) $SiCl_4$
(c) CaSe
(d) KNO_3
(e) $Al_2(SO_4)_3$
(f) $Ni(HCO_3)_2$
(g) $NaHSO_4$
(h) $(NH_4)_2Cr_2O_7$
(i) $Ca(C_2H_3O_2)_2$
(j) $Sr(OH)_2$

4.95
(a) strontium chloride
(b) calcium nitrate
(c) copper(II) sulfide
(d) tin(II) phosphate
(e) nickel(II) chlorate
(f) zinc acetate
(g) bromic acid
(h) mercury(II) bromide
(i) cobalt(II) sulfate
(j) potassium dihydrogen arsenate

4.97 (a) $Pb(C_2H_3O_2)_4$ (b) Na_2Se (c) $Ba_3(PO_4)_2$
(d) HI (e) $HI(aq)$ (f) PBr_3
(g) $Ca(OCl)_2$ [or $Ca(ClO)_2$] (h) $Ag_2C_2O_4$ (i) H_2CrO_4
(j) SiF_4

4.99 (a) S_3O_9 (b) ICl_5 (c) $Cr_2(SO_4)_3$ (d) $Fe_2(SO_4)_3$
(e) PbS (f) HIO_4 (g) $LiOI$ [or $LiIO$] (h) $Hg(NO_3)_2$
(i) $Au_2(SO_4)_3$ (j) Bi_2O_5

4.101 (a) $Fe_2(SO_4)_3$ (b) $FeCl_2$ (c) $Hg_2(NO_3)_2$
(d) $CuCl$ (e) $SnCl_4$ (f) $Co(OH)_2$
(g) $AuCl_3$ (h) $Cr(C_2H_3O_2)_3$

5 CHEMICAL REACTIONS IN AQUEOUS SOLUTION

5.1 Solutions are usually employed for carrying out chemical reactions because the homogeneous nature of solutions allows dissolved substances to intermingle freely. Thus, the reactions involving solutions are permitted to occur very rapidly.

5.3 Solubility: the amount of solute required to produce a saturated solution with a given amount of solvent at a particular temperature

5.5 Yes, for a solute which has a limited solubility in a particular solvent, a saturated solution will have a relatively small proportion of solute in large amounts of solvent.

5.7 To prepare a supersaturated solution of sugar in water, heat the water and add enough sugar to form a saturated solution at the elevated temperature. When the solution is allowed to cool, a supersaturated solution is the result if no crystals are present for the excess to crystallize onto.

5.9 An electrolyte is a substance which when dissolved in a solvent produces ions which make the solution able to conduct electricity. A nonelectrolyte is a substance which, when dissolved in a solvent, does not produce ions; therefore, it does not give an electrically conducting solution.

5.11 $KCl(aq) \rightarrow K^+(aq) + Cl^-(aq)$

$(NH_4)_2SO_4(aq) \rightarrow 2NH_4^+(aq) + SO_4^{2-}(aq)$

$Na_3PO_4(aq) \rightarrow 3Na^+(aq) + PO_4^{3-}(aq)$

$NaOH(aq) \rightarrow Na^+(aq) + OH^-(aq)$

$HCl(aq) \rightarrow H^+(aq) + Cl^-(aq)$

5.13 A dynamic equilibrium is one in which opposing processes occur at equal rates, so there is no net change in the system, e.g., ions react to form molecules while molecules react to form ions.

5.15 $H_2O \rightleftharpoons H^+ + OH^-$ (or, $2H_2O \rightleftharpoons H_3O^+ + OH^-$)

5.17 The position of the equilibrium lies mainly to the left in favor of undissociated H_2O in the equation for the dissociation of water. For HCl the position of equilibrium lies mainly to the right, virtually 100% in favor of the ionic products.

5.19 Precipitate

5.21 A spectator ion is an ion that does not change during a reaction.

5.23 In a molecular equation all reactants and products are written as if they were molecules. An ionic equation more accurately represents a reaction as it actually occurs in solution by showing all soluble ionic substances as being dissociated. A net ionic equation only represents the net chemical change that occurs. The spectator ions are not shown in a net ionic equation.

5.25 Ionic equation:

$Ca^{2+}(aq) + 2Cl^{-}(aq) + 2K^{+}(aq) + CO_3^{2-}(aq) \rightarrow CaCO_3(s) + 2K^{+}(aq) + 2Cl^{-}(aq)$

Net ionic equation: $Ca^{2+}(aq) + CO_3^{2-}(aq) \rightarrow CaCO_3(s)$

5.27 (a) $Cu^{2+} + 2NO_3^{-} + 2Na^{+} + 2OH^{-} \rightarrow Cu(OH)_2(s) + 2Na^{+} + 2NO_3^{-}$

$Cu^{2+}(aq) + 2OH^{-}(aq) \rightarrow Cu(OH)_2(s)$ (net ionic)

(b) $3Ba^{2+} + 6Cl^{-} + 2Al^{3+} + 3SO_4^{2-} \rightarrow 3BaSO_4(s) + 2Al^{3+} + 6Cl^{-}$

$Ba^{2+}(aq) + SO_4^{2-}(aq) \rightarrow BaSO_4(s)$ (net ionic)

(c) $Hg_2^{2+} + 2NO_3^{-} + 2H^{+} + 2Cl^{-} \rightarrow Hg_2Cl_2(s) + 2H^{+} + 2NO_3^{-}$

$Hg_2^{2+}(aq) + 2Cl^{-}(aq) \rightarrow Hg_2Cl_2(s)$ (net ionic)

(d) $2Bi^{3+} + 6NO_3^{-} + 6Na^{+} + 3S^{2-} \rightarrow Bi_2S_3(s) + 6Na^{+} + 6NO_3^{-}$

$2Bi^{3+}(aq) + 3S^{2-}(aq) \rightarrow Bi_2S_3(s)$ (net ionic)

(e) $Ca^{2+} + 2Cl^{-} + 2Na^{+} + SO_4^{2-} \rightarrow CaSO_4(s) + 2Na^{+} + 2Cl^{-}$

$Ca^{2+}(aq) + SO_4^{2-}(aq) \rightarrow CaSO_4(s)$ (net ionic)

5.29 An acid is any substance that increases the concentration of hydronium ions (hydrogen ions) by reaction with water. A base is any substance that increases the concentration of hydroxide ions in aqueous solutions.

5.31 A strong acid is essentially 100% ionized in solution while only a small fraction of a weak acid ionizes in solution.

5.33 (a) $H_2SO_3 \rightleftharpoons H^{+} + HSO_3^{-}$

$HSO_3^{-} \rightleftharpoons H^{+} + SO_3^{2-}$

(b) $H_3AsO_4 \rightleftharpoons H^{+} + H_2AsO_4^{-}$

$H_2AsO_4^{-} \rightleftharpoons H^{+} + HAsO_4^{2-}$

$HAsO_4^{2-} \rightleftharpoons H^{+} + AsO_4^{3-}$

5.35 Acid anhydride - nonmetal oxides that react with H_2O to yield acid solutions.
e.g., $SO_2 + H_2O \rightleftharpoons H_2SO_3$

Basic anhydride - metal oxides that react with H_2O to give corresponding hydroxides. e.g., $CaO + H_2O \rightarrow Ca(OH)_2$

5.37 $H_2O + N_2O_5 \rightarrow 2HNO_3$

5.39 A base turns litmus blue. Therefore, the element would be classified as a metal because its oxide yields a base on reaction with water while if it were a nonmetal its oxide would be an acidic anhydride.

5.41 $O^{2-} + H_2O \rightarrow 2OH^-$

5.43 $NH_3(aq) + H_2O \rightleftharpoons NH_4^+(aq) + OH^-(aq)$ Ammonia is a weak base.

5.45 $H_3O^+ + OH^- \rightarrow 2H_2O$ or $H^+ + OH^- \rightarrow H_2O$

5.47 Acid salts are the products of partial neutralization of a polyprotic acid. Examples include $NaHCO_3$, $NaHSO_4$, and Na_2HPO_4.

KH_2PO_4	potassium dihydrogen phosphate
K_2HPO_4	dipotassium hydrogen phosphate
K_3PO_4	tripotassium phosphate (or potassium phosphate)

5.49

Soluble	Insoluble
KCl	$PbSO_4$
$(NH_4)_2SO_4$	$Mn(OH)_2$
$AgNO_3$	$FePO_4$
$Zn(ClO_4)_2$	$CaCO_3$
$Ba(C_2H_3O_2)_2$	NiO

5.51 (a) Ionic: $Al(OH)_3(s) + 3H^+(aq) + 3Cl^-(aq) \rightarrow Al^{3+}(aq) + 3Cl^-(aq) + 3H_2O$

Net ionic: $Al(OH)_3(s) + 3H^+(aq) \rightarrow Al^{3+}(aq) + 3H_2O$

(b) Ionic: $CuCO_3(s) + 2H^+(aq) + SO_4^{2-}(aq) \rightarrow Cu^{2+}(aq) + SO_4^{2-}(aq) + H_2O + CO_2(g)$

Net ionic: $CuCO_3(s) + 2H^+(aq) \rightarrow Cu^{2+}(aq) + H_2O + CO_2(g)$

(c) Ionic: $Cr_2(CO_3)_3(s) + 6H^+(aq) + 6NO_3^-(aq)$

$\rightarrow 2Cr^{3+}(aq) + 6NO_3^-(aq) + 3H_2O + 3CO_2(g)$

Net ionic: $Cr_2(CO_3)_3(s) + 6H^+(aq) \rightarrow 2Cr^{3+}(aq) + 3H_2O + 3CO_2(g)$

5.53 (a) $CoS(s) + 2H^+(aq) \rightarrow H_2S(g) + Co^{2+}(aq)$

(b) $PbCO_3(s) + 2H^+(aq) \rightarrow H_2O + CO_2(g) + Pb^{2+}(aq)$

(c) $PbCO_3(s) + 2H^+(aq) + SO_4^{2-}(aq) \rightarrow PbSO_4(s) + H_2O + CO_2(g)$

(d) $Sn^{2+}(aq) + 2OH^-(aq) \rightarrow Sn(OH)_2(s)$

(e) $Ag_2O(s) + 2H^+(aq) + 2Cl^-(aq) \rightarrow 2AgCl(s) + H_2O$

(f) (This reaction does not have a driving force.)

5.55 (a) no reaction between reactants

(b) no reaction between reactants

(c) $K_2S(aq) + Ni(C_2H_3O_2)_2(aq) \rightarrow 2KC_2H_3O_2(aq) + NiS(s)$

$2K^+(aq) + S^{2-}(aq) + Ni^{2+}(aq) + 2C_2H_3O_2^-(aq) \rightarrow 2K^+(aq) + 2C_2H_3O_2^-(aq) + NiS(s)$

$Ni^{2+}(aq) + S^{2-}(aq) \rightarrow NiS(s)$

(d) $MgSO_4(aq) + 2LiOH(aq) \rightarrow Li_2SO_4(aq) + Mg(OH)_2(s)$

$Mg^{2+}(aq) + SO_4^{2-}(aq) + 2Li^+(aq) + 2OH^-(aq) \rightarrow 2Li^+(aq) + SO_4^{2-}(aq) + Mg(OH)_2(s)$

$Mg^{2+}(aq) + 2OH^-(aq) \rightarrow Mg(OH)_2(s)$

(e) $AgC_2H_3O_2(aq) + KCl(aq) \rightarrow AgCl(s) + KC_2H_3O_2(aq)$

$Ag^+(aq) + C_2H_3O_2^-(aq) + K^+(aq) + Cl^-(aq) \rightarrow AgCl(s) + K^+(aq) + C_2H_3O_2^-(aq)$

$Ag^+(aq) + Cl^-(aq) \rightarrow AgCl(s)$

5.57 $H_2O + CO_2(g) \rightarrow H_2CO_3(aq)$

$H_2CO_3(aq) + 2NaOH(aq)$ (excess) $\rightarrow Na_2CO_3(aq) + 2H_2O$

5.59 The following are only the answers. The necessary half-reactions and steps are not shown. The states of the reactants and products are also not shown.

(a) $8H^+ + Cr_2O_7^{2-} + 3CH_3CH_2OH \rightarrow 2Cr^{3+} + 3CH_3CHO + 7H_2O$

(b) $4H^+ + PbO_2 + 2Cl^- \rightarrow Pb^{2+} + Cl_2 + 2H_2O$

(c) $14H^+ + 2Mn^{2+} + 5BiO_3^- \rightarrow 2MnO_4^- + 5Bi^{3+} + 7H_2O$

(d) $3H_2O + ClO_3^- + 3HAsO_2 \rightarrow 3H_3AsO_4 + Cl^-$

(e) $2H_2O + PH_3 + 2I_2 \rightarrow H_3PO_2 + 4I^- + 4H^+$

(f) $16H^+ + 2MnO_4^- + 10S_2O_3^{2-} \rightarrow 5S_4O_6^{2-} + 2Mn^{2+} + 8H_2O$

(g) $4H^+ + 2Mn^{2+} + 5PbO_2 \rightarrow 2MnO_4^- + 5Pb^{2+} + 2H_2O$

(h) $2H^+ + As_2O_3 + 2NO_3^- + 2H_2O \rightarrow 2H_3AsO_4 + N_2O_3$

(i) $8H_2O + 2P + 5Cu^{2+} \rightarrow 5Cu + 2H_2PO_4^- + 12H^+$

(j) $6H^+ + 2MnO_4^- + 5H_2S \rightarrow 2Mn^{2+} + 5S + 8H_2O$

5.61 (See the note at the beginning of the answers to question 5.59).

(a) $3H_2O + P_4 + 3OH^- \rightarrow PH_3 + 3H_2PO_2^-$

(b) $12H^+ + 12Cu + 12Cl^- + As_4O_6 \rightarrow 12CuCl + 4As + 6H_2O$

(c) $9H_2O + 5IPO_4 \rightarrow I_2 + 3IO_3^- + 5H_2PO_4^- + 8H^+$

(d) $3NO_2 + H_2O \rightarrow 2NO_3^- + NO + 2H^+$

(e) $6OH^- + 3Br_2 \rightarrow 5Br^- + BrO_3^- + 3H_2O$

(f) $4HSO_2NH_2 + 6NO_3^- \rightarrow 4SO_4^{2-} + 2H^+ + 5N_2O + 5H_2O$

(g) $4H^+ + 2ClO_3^- + 2Cl^- \rightarrow 2ClO_2 + Cl_2 + 2H_2O$

(h) $2OH^- + 2ClO_2 \rightarrow ClO_2^- + ClO_3^- + H_2O$

(i) $6OH^- + 3Se \rightarrow 2Se^{2-} + SeO_3^{2-} + 3H_2O$

(j) $3H_2O + 5ICl \rightarrow 2I_2 + IO_3^- + 5Cl^- + 6H^+$

(k) $4OH^- + 2FNO_3 \rightarrow O_2 + 2F^- + 2NO_3^- + 2H_2O$

(l) $2H_2O + 4Fe(OH)_2 + O_2 \rightarrow 4Fe(OH)_3$

5.63 (a) CrO_4^{2-} --- yellow

(b) $Cr_2O_7^{2-}$ --red-orange

(c) MnO_4^- --- purple

5.65 $3HSO_3^-(aq) + Cr_2O_7^{2-}(aq) + 5H^+(aq) \rightarrow 3SO_4^{2-}(aq) + 2Cr^{3+}(aq) + 4H_2O$
or (if very acidic)
$3H_2SO_3(aq) + Cr_2O_7^{2-}(aq) + 2H^+(aq) \rightarrow 3SO_4^{2-}(aq) + 2Cr^{3+}(aq) + 4H_2O$

5.67 $3SO_3^{2-}(aq) + 2CrO_4^{2-}(aq) + H_2O \rightarrow 3SO_4^{2-}(aq) + 2CrO_2^-(aq) + 2OH^-(aq)$

5.69 Parts per million is the number of grams or volumes of solute per million (10^6) grams or volumes of solution.

5.71 (a) $\frac{0.001 \text{ g F}^-}{1{,}000 \text{ g soln.}} \times 100 = \mathbf{1 \times 10^{-4}\ \%\ F^-\ by\ mass}$

(b) $\frac{0.001 \text{ g F}^-}{1{,}000 \text{ g soln.}} \times 1{,}000{,}000 = \mathbf{1\ part\ F^-\ per\ million}$

(c) $\frac{0.001 \text{ g F}^-}{1{,}000 \text{ g soln.}} \times 1{,}000{,}000{,}000 = \mathbf{1 \times 10^3\ parts\ F^-\ per\ billion}$

5.73 (a) $\frac{1.50 \text{ mol NaCl}}{2.00 \text{ L soln.}} = \mathbf{0.750\ M}$

(b) **0.992 M**

(c) **0.556 M**

(d) $\frac{85.5 \text{ g HNO}_3}{1.00 \text{ L soln.}} \times \frac{1 \text{ mol HNO}_3}{63.02 \text{ g HNO}_3} = \mathbf{1.36\ M}$

(e) $\frac{44.5 \text{ g NH}_4\text{C}_2\text{H}_3\text{O}_2}{600 \text{ mL soln.}} \times \frac{1{,}000 \text{ mL}}{\text{L}} \times \frac{1 \text{ mol NH}_4\text{C}_2\text{H}_3\text{O}_2}{77.08 \text{ NH}_4\text{C}_2\text{H}_3\text{O}_2} = \mathbf{0.962\ M}$

5.75 $$\frac{0.150 \text{ mol } Na_2CO_3}{\text{L soln.}} \text{ x } 0.300 \text{ L soln. x } \frac{106.0 \text{ g } Na_2CO_3}{\text{mol}} = \mathbf{4.77 \text{ g } Na_2CO_3}$$

5.77 In pure nitric acid, nitric acid is both the solute and the solution. Therefore, a density of 1.513 g/mL can be expressed as 1.513 g solute/mL solution.

$$\frac{1.513 \text{ g } HNO_3}{\text{mL soln.}} \text{ x } \frac{1{,}000 \text{ mL soln.}}{\text{L soln.}} \text{ x } \frac{1 \text{ mole } HNO_3}{63.012 \text{ g } HNO_3} = \mathbf{24.01 \text{ M}}$$

5.79 (a) **0.100 M Li^+** and **0.100 M Cl^-**
(b) **0.250 M Ca^{2+}** and **0.500 M Cl^-**
(c) **2.40 M NH_4^+** and **1.20 M SO_4^{2-}**
(d) **0.600 M Na^+** and ~ **0.600 M HSO_4^-**
(e) **0.800 M Fe^{3+}** and **1.20 M SO_4^{2-}**

5.81 $$\frac{0.100 \text{ mol } SO_4^{2-}}{\text{L}} \text{ x } \frac{1 \text{ mol } Na_2SO_4}{1 \text{ mol } SO_4^{2-}} \Leftrightarrow \frac{0.100 \text{ mol } Na_2SO_4}{\text{L}} \Leftrightarrow \mathbf{0.100 \text{ M } Na_2SO_4}$$

5.83 **0.0700 M** (See problem 5.81 for an example of the method for obtaining this answer).

5.85 M.M. of $CuSO_4 \cdot 5H_2O$ = 249.68 The concentration of the solution is:

$$\frac{10.45 \text{ g salt}}{150.0 \text{ mL soln.}} \text{ x } \frac{1{,}000 \text{ mL}}{\text{L}} \text{ x } \frac{1 \text{ mol salt}}{249.68 \text{ g}} = 0.2790 \text{ M salt}$$

The salt solution will contain: **0.2790 M Cu^{2+}** and **0.2790 M SO_4^{2-}**

5.87 $$3.22 \text{ g Cu} \times \frac{1 \text{ mol Cu}}{63.546 \text{ g Cu}} \times \frac{8 \text{ mol } HNO_3}{3 \text{ mol Cu}} \times \frac{1 \text{ L soln.}}{1.250 \text{ mol } HNO_3} \Leftrightarrow$$

0.1081 L soln. = **108 mL** (based upon H^+ provided by the HNO_3)

$$3.22 \text{ g Cu} \times \frac{1 \text{ mol Cu}}{63.546 \text{ g Cu}} \times \frac{2 \text{ mol } HNO_3}{3 \text{ mol Cu}} \times \frac{1 \text{ L soln.}}{1.250 \text{ mol } HNO_3} \Leftrightarrow$$

0.0270 L soln. = **27.0 mL** (based upon NO_3^- provided by the HNO_3)

5.89 (a) $H_3PO_4(aq) + 3NaOH(aq) \rightarrow Na_3PO_4(aq) + 3H_2O$

$$\frac{0.170 \text{ mol } H_3PO_4}{\text{L soln.}} \times 0.500 \text{ L soln.} \times \frac{3 \text{ mol NaOH}}{\text{mol } H_3PO_4} \times \frac{1 \text{ L NaOH soln.}}{0.300 \text{ mol NaOH}} \Leftrightarrow$$

0.850 L NaOH soln. = **850 mL**

(b) $H_3PO_4(aq) + 2NaOH(aq) \rightarrow Na_2HPO_4(aq) + 2H_2O$

$$\frac{0.170 \text{ mol } H_3PO_4}{\text{L soln.}} \times 0.500 \text{ L soln.} \times \frac{2 \text{ mol NaOH}}{\text{mol } H_3PO_4} \times \frac{1 \text{ L NaOH soln.}}{0.300 \text{ mol NaOH}} \Leftrightarrow$$

0.567 L NaOH soln. = **567 mL**

(c) $H_3PO_4(aq) + NaOH(aq) \rightarrow NaH_2PO_4(aq) + H_2O$

$$\frac{0.170}{1} \times \frac{.500}{1} \times \frac{1}{1} \times \frac{1}{0.300} = 0.283 \text{ L} = \mathbf{283 \text{ mL}}$$

5.91 $3BaCl_2(aq) + Fe_2(SO_4)_3(aq) \rightarrow 3BaSO_4(s) + 2FeCl_3(aq)$

$$\frac{0.200 \text{ mol } Fe_2(SO_4)_3}{\text{L}} \times 0.0250 \text{ L} \times \frac{3 \text{ mol } BaCl_2}{\text{mol } Fe_2(SO_4)_3} \times \frac{1 \text{ L } BaCl_2 \text{ soln.}}{0.100 \text{ mol } BaCl_2}$$

$$\times \frac{1{,}000 \text{ mL}}{\text{L}} \Leftrightarrow \mathbf{150 \text{ mL } BaCl_2 \text{ soln.}}$$

5.93 (a) Molecular equation: $AgNO_3(aq) + NaCl(aq) \rightarrow NaNO_3(aq) + AgCl(s)$
Net ionic equation: $\mathbf{Ag^+(aq) + Cl^-(aq) \rightarrow AgCl(s)}$

(b) 20.0 mL of 0.200 M $AgNO_3$ contains:

$$0.0200 \text{ L x } \frac{0.200 \text{ mol}}{\text{L}} \text{ or } 0.00400 \text{ mole of } Ag^+$$

30.0 mL of 0.200 M NaCl contains:

$$0.0300 \text{ L x } \frac{0.200 \text{ mol}}{\text{L}} \text{ or } 0.00600 \text{ mole of } Cl^-$$

From this, one can see that the Ag^+ is the limiting reactant and that only **0.00400 mole of AgCl can be precipitated.**

(c) $$0.00400 \text{ mol AgCl x } \frac{143.3 \text{ g AgCl}}{\text{mol AgCl}} = \mathbf{0.573 \text{ g AgCl}}$$

(d) The amount of each ion before reaction is: Ag^+ = 0.00400 moles, NO_3^- = 0.00400 moles, Na^+ = 0.00600 moles, and Cl^- = 0.00600 moles. The precipitation process will remove 0.00400 moles of Ag^+ and 0.00400 moles of Cl^- leaving in solution 0.0 moles Ag^+, 0.00200 moles Cl^-, 0.00400 moles NO_3^- and 0.00600 moles Na^+. The concentration of each ion will be the number of moles of ion in the final solution divided by the total volume of solution.

$$Ag^+ = \mathbf{0 \text{ M}} \qquad Cl^- = \frac{0.00200 \text{ mol}}{0.0500 \text{ L}} = \mathbf{0.0400 \text{ M}}$$

$$NO_3^- = \frac{0.00400 \text{ mol}}{0.0500 \text{ L}} = \mathbf{0.0800 \text{ M}} \qquad Na^+ = \frac{0.00600 \text{ mol}}{0.0500 \text{ L}} = \mathbf{0.120 \text{ M}}$$

5.95 (a) $$\frac{0.0500 \text{ mol NaOH}}{\text{L NaOH soln.}} \text{ x } 0.0172 \text{ L NaOH soln. x } \frac{1 \text{ mol cap. acid}}{1 \text{ mol NaOH}}$$

$$\Leftrightarrow 0.000860 \text{ mol caproic acid}$$

$$\frac{0.100 \text{ g cap. acid}}{0.000860 \text{ mol}} = \mathbf{116 \text{ g/mol} = \text{M.M.}}$$

(b) C_3H_6O empirical formula mass = 58.1

From its molecular mass and its empirical formula mass, its molecular formula must be twice its empirical formula. $\mathbf{C_6H_{12}O_2}$

5.97 (a) $\mathbf{3Ba^{2+}(aq)+6OH^-(aq)+2Al^{3+}(aq)+3SO_4^{2-}(aq)\rightarrow 2Al(OH)_3(s)+3BaSO_4(s)}$

(b) $0.270 \text{ mol } Ba^{2+} L^{-1} \times 0.0400 \text{ L} = 0.0108 \text{ mol } Ba^{2+}$

$$0.0108 \text{ mol } Ba^{2+} \times \frac{2 \text{ mol } OH^-}{\text{mol } Ba^{2+}} \Leftrightarrow 0.0216 \text{ mol } OH^-$$

$$\frac{0.330 \text{ mol } Al_2(SO_4)_3}{\text{L}} \times \frac{2 \text{ mol } Al^{3+}}{\text{mol } Al_2(SO_4)_3} \times 0.0250 \text{ L} \Leftrightarrow 0.0165 \text{ mol } Al^{3+}$$

$$\frac{0.330 \text{ mol } Al_2(SO_4)_3}{\text{L}} \times \frac{3 \text{ mol } SO_4^{2-}}{\text{mol } Al_2(SO_4)_3} \times 0.0250 \text{ L} \Leftrightarrow 0.0248 \text{ mol } SO_4^{2-}$$

Ba^{2+} is the limiting reactant for the formation of $BaSO_4$.

$$0.0108 \text{ mol } Ba^{2+} \times \frac{3 \text{ mol } BaSO_4}{3 \text{ mol } Ba^{2+}} \times \frac{233.4 \text{ g } BaSO_4}{\text{mol } BaSO_4} \Leftrightarrow 2.52 \text{ g of } BaSO_4 \text{ ppt.}$$

OH^- is the limiting reactant for the formation of $Al(OH)_3$

$$0.0216 \text{ mol } OH^- \times \frac{2 \text{ mol } Al(OH)_3}{6 \text{ mol } OH^-} \times \frac{78.0 \text{ g } Al(OH)_3}{\text{mol } Al(OH)_3} \Leftrightarrow 0.562 \text{ g of } Al(OH)_3 \text{ ppt.}$$

Total weight of ppt. is 2.52 + 0.562 = **3.08 g**

(c) **~0 M Ba^{2+}, ~0 M OH^-**

$$0.0108 \text{ mol } Ba^{2+} \times \frac{1 \text{ mol } SO_4^{2-}}{\text{mol } Ba^{2+}} \Leftrightarrow 0.0108 \text{ mol } SO_4^{2-} \text{ reacted}$$

$$\frac{0.0248 \text{ mol } SO_4^{2-} \text{ total} - 0.0108 \text{ mol } SO_4^{2-} \text{ ppt.}}{0.0650 \text{ L}} = \mathbf{0.215\ M\ SO_4^{2-}}$$

$$0.0216 \text{ mol } OH^- \times \frac{\text{mol } Al^{3+}}{3 \text{ mol } OH^-} \Leftrightarrow 0.00720 \text{ mol } Al^{3+} \text{ reacted}$$

$$\frac{0.0165 \text{ mol } Al^{3+} - 0.00720 \text{ mol } Al^{3+} \text{ ppt.}}{0.0650 \text{ L}} = \mathbf{0.143\ M\ Al^{3+}}$$

5.99 $0.694 \text{ g AgCl} \times \frac{1 \text{ mol AgCl}}{143.4 \text{ g}} \times \frac{1 \text{ mol Cl}}{\text{mol AgCl}} \Leftrightarrow 0.00484 \text{ mol Cl}$

$0.00484 \text{ mol Cl} \times \frac{35.45 \text{ g Cl}}{\text{mol}} = 0.172 \text{ g Cl}$

g Ti = 0.249 g of sample - 0.172 g Cl = 0.077 g Ti

$0.077 \text{ g Ti} \times \frac{1 \text{ mol Ti}}{47.9 \text{ g}} = 0.0016 \text{ mol Ti}$

Formula is: $Ti_{0.0016}Cl_{0.00484}$ or $Ti_{0.0016/0.0016}Cl_{0.00484/0.0016}$ or $\mathbf{TiCl_3}$

5.101 $2AgCl(s)(excess) + CuBr_2(aq) \rightarrow 2AgBr(s) + CuCl_2(aq) + AgCl(s)$

$1.800 \text{ g AgCl(initial)} \times \frac{1 \text{ mol AgCl}}{143.32 \text{ g AgCl}} = 0.01256 \text{ mol AgCl(initial)}$

0.01256 mol AgCl(initial) = mol AgBr + mol AgCl(excess)
mol AgBr = X
mol AgCl(final) = 0.01256 - X

2.052 g sample = (X mol AgBr x 187.77 g/mol) +
[(0.01256 - X) mol AgCl x 143.32 g/mol]

2.052 = 187.77X + 1.800 - 143.32X X = 0.00567 mol AgBr

$0.00567 \text{ mol AgBr} \times \frac{1 \text{ mol } CuBr_2}{2 \text{ mol AgBr}} \times \frac{223.35 \text{ g } CuBr_2}{\text{mol } CuBr_2} \Leftrightarrow 0.633 \text{ g } CuBr_2$

$\frac{0.633 \text{ g } CuBr_2}{1.850} \times 100\% = \mathbf{34.2\%}$

5.103 An indicator signals when the reaction is complete.
(a) colorless

(b) pink

5.105 $CaCO_3 \rightarrow CaO \rightarrow Ca(OH)_2$

$Ca(OH)_2(aq) + 2HCl(aq) \rightarrow CaCl_2(aq) + 2H_2O$

$$\frac{0.120 \text{ mol HCl}}{\text{L}} \times 0.03725 \text{ L} \times \frac{1 \text{ mol Ca(OH)}_2}{2 \text{ mol HCl}} \Leftrightarrow 0.00224 \text{ mol Ca(OH)}_2$$

$$0.00224 \text{ mol Ca(OH)}_2 = 0.00224 \text{ mol CaCO}_3$$

$$\frac{0.00224 \text{ mol CaCO}_3 \times 100.1 \text{ g mol}^{-1}}{1.030 \text{ g sample}} \times 100\% = \mathbf{21.8\%}$$

5.107 (a) $$\frac{0.05000 \text{ mol HCl}}{\text{L}} \times 0.0500 \text{ L} = 0.00250 \text{ mol HCl total}$$

$$\frac{0.0600 \text{ mol NaOH}}{\text{L}} \times 0.03057 \text{ L} \times \frac{1 \text{ mol HCl neut.}}{1 \text{ mol NaOH}}$$

$$\Leftrightarrow 0.00183 \text{ mol HCl neut. by NaOH}$$

0.00250 mol HCl total = 0.00183 mol HCl neut. by NaOH + X mol HCl neut. by NH_3

X = 0.00067 mol HCl neutralized by NH_3

(b) $$0.00067 \text{ mol HCl} \times \frac{1 \text{ mol NH}_3}{\text{mol HCl}} \times \frac{1 \text{ mol N}}{\text{mol NH}_3} \times \frac{14.01 \text{ g N}}{\text{mol N}} \Leftrightarrow \mathbf{0.0094 \text{ g N}}$$

(c) $$\frac{0.0094 \text{ g N}}{0.0500 \text{ g sample}} \times 100\% = \mathbf{19\% \text{ N in sample}}$$

$$\frac{14.01 \text{ g N in gly}}{75.08 \text{ g gly}} \times 100\% = \mathbf{18.66\% \text{ N in gly}}$$

Glycine and the sample have the same percent nitrogen to two significant figures.

5.109 The number of equivalents of A that react is exactly equal to the number of equivalents of B that react in any given reaction.

5.111 $5.00 \text{ eq } H_3PO_4 \times \frac{1 \text{ mol } H_3PO_4}{3 \text{ eq } H_3PO_4} = \mathbf{1.67 \text{ mol } H_3PO_4}$

5.113 (a) $MnSO_4 \rightarrow Mn_2O_3$ or $Mn^{2+} \rightarrow Mn^{3+}$ (1 electron change)

F.M. of $MnSO_4$= 151.00 $\quad$ $\text{eq wt} = \frac{151.00 \text{ g}}{\text{mol}} \times \frac{1 \text{ mol}}{\text{eq}} = \mathbf{151.00 \text{ g/eq}}$

(b) $Mn^{2+} \rightarrow Mn^{4+}$ $\quad$ $\text{eq wt} = \frac{151.00 \text{ g}}{\text{mol}} \times \frac{1 \text{ mol}}{2 \text{ eq}} = \mathbf{75.50 \text{ g/eq}}$

(c) $Mn^{2+} \rightarrow Mn^{6+}$ $\quad$ $\text{eq wt} = \frac{151.00 \text{ g}}{\text{mol}} \times \frac{1 \text{ mol}}{4 \text{ eq}} = \mathbf{37.75 \text{ g/eq}}$

(d) $Mn^{2+} \rightarrow Mn^{7+}$ $\quad$ $\text{eq wt} = \frac{151.00 \text{ g}}{\text{mol}} \times \frac{1 \text{ mol}}{5 \text{ eq}} = \mathbf{30.20 \text{ g/eq}}$

5.115 $Mn^{2+} \rightarrow Mn^{7+}$

$\frac{0.100 \text{ eq}}{\text{L}} \times 0.300 \text{ L} \times \frac{1 \text{ mol}}{5 \text{ eq}} \times \frac{259.1 \text{ g}}{\text{mol}} = \mathbf{1.55 \text{ g } MnSO_4 \cdot 6H_2O}$

5.117 (a) $\frac{98.00 \text{ g } H_3PO_4}{\text{mol}} \times \frac{1 \text{ mol}}{2 \text{ eq}} = \mathbf{49.00 \text{ g } H_3PO_4/eq}$

(b) $\frac{100.5 \text{ g } HClO_4}{\text{mol}} \times \frac{1 \text{ mol}}{\text{eq}} = \mathbf{100.5 \text{ g } HClO_4/eq}$

(c) $I^{5+} \rightarrow I^-$ $(-6e^-)$ $\quad$ $\frac{197.89 \text{ g } NaIO_3}{\text{mol}} \times \frac{1 \text{ mol}}{6 \text{ eq}} = \mathbf{32.98 \text{ g } NaIO_3/eq}$

(d) $I^{5+} \rightarrow I^0$ $(-5e^-)$ $\quad$ $\frac{197.89 \text{ g } NaIO_3}{\text{mol}} \times \frac{1 \text{ mol}}{5 \text{ eq}} = \mathbf{39.58 \text{ g } NaIO_3/eq}$

(e) $\frac{78.01 \text{ g } Al(OH)_3}{\text{mol}} \times \frac{1 \text{ mol}}{3 \text{ eq}} = \mathbf{26.00 \text{ g } Al(OH)_3/eq}$

5.119 $\frac{0.850 \text{ eq Ba(OH)}_2}{\text{L}} \times 0.129 \text{ L} \times \frac{1 \text{ eq acid}}{1 \text{ eq Ba(OH)}_2} = 0.110 \text{ equivalents acid}$

$\frac{4.93 \text{ g acid}}{0.110 \text{ eq acid}} = \mathbf{44.8\ g/eq}$

5.121 Use $V_AN_A = V_BN_B$ and $V_{dil}N_{dil} = V_{conc}N_{conc}$

For neutralization: 41.0 mL(B) x 0.255 N(B) = 5.00 mL(A) x ?N(A)

N(A) = 2.09 N HCl (the dil. HCl)

From dilution: 50.0 mL x 2.09 N = 10.0 mL x ?N

N = 10.45 To 3 significant figures the answer is **10.5 N HCl or 10.5 M HCl**

5.123 Use $V_{conc}N_{conc} = V_{dil}N_{dil}$

85.0 mL x 1.00 = V_{dil} x 0.650 N

V_{dil} = 131 mL

H_2O added = 131 mL - 85.0 mL = **46 mL**

5.125 $\frac{825 \text{ g solute}}{10^6 \text{ g solution}} \times 100 = 8.25 \times 10^{-2}\% = \mathbf{0.0825\%\ by\ mass}$

$\frac{825 \text{ g benzene}}{1.00 \times 10^6 \text{ g solution}} \times \frac{1{,}000 \text{ g}}{1{,}000 \text{ mL}} \times \frac{1{,}000 \text{ mL}}{\text{L}} \times \frac{1 \text{ mol benzene}}{78.11 \text{ g}} = \mathbf{0.0106\ M}$

5.127 $5Fe^{2+} + MnO_4^- + 8H^+ \rightarrow 5Fe^{3+} + Mn^{2+} + 4H_2O$

$$\frac{0.00400 \text{ mol } MnO_4^-}{L} \times 0.0158 \text{ L} \times \frac{5 \text{ mol } Fe^{2+}}{\text{mol } MnO_4^-}$$

$$\times \frac{1 \text{ mol } FeSO_4}{\text{mol } Fe^{2+}} \times \frac{151.9 \text{ g } FeSO_4}{\text{mol } FeSO_4} \Leftrightarrow 0.0480 \text{ g } FeSO_4$$

$$\frac{0.0480 \text{ g } FeSO_4}{0.1000 \text{ g sample}} \times 100\% = \mathbf{48.0\%\ FeSO_4\ by\ mass}$$

5.129 $\frac{1.00 \text{ mol NaOH}}{L} \times 19.6 \times 10^{-3} \text{L} \times \frac{1 \text{ mol HCl excess}}{1 \text{ mol NaOH}} \Leftrightarrow 0.0196 \text{ mol HCl excess}$

$$\frac{1.00 \text{ mol HCl}}{L} \times 0.100 \text{ L HCl} = 0.100 \text{ mol HCl total}$$

0.100 mol HCl total - 0.0196 mol HCl excess = 0.0804 mol HCl reacted

$$0.0804 \text{ mol HCl react.} \times \frac{1 \text{ mol (CaO + MgO)}}{2 \text{ mol HCl}} \Leftrightarrow 0.0402 \text{ mol (CaO + MgO)}$$

Let X = moles CaO and 0.0402 - X = moles of MgO.

Then:
$2.000 \text{ g} = (X)(56.08 \text{ g/mol}) + (0.0402 - X)(40.31)$
$2.000 = 56.08X + 1.620 - 40.31X$
$0.380 = 15.77X$

X = 0.024 moles CaO = moles $CaCO_3$
0.016 = moles MgO = moles $MgCO_3$

$$\% \ CaCO_3 = \frac{0.024 \text{ mol } CaCO_3 \times 100.1 \text{ g/mol}}{(0.024 \text{ mol } CaCO_3 \times 100.1 \text{ g/mol}) + (0.016 \text{ mol } MgCO_3 \times 84.31 \text{ g/mol}}$$

$$\times 100\% = \frac{2.40 \times 100}{2.40 + 1.3} = \mathbf{65\%\ CaCO_3}$$

$$\% \ MgCO_3 = \frac{1.3}{2.40 \times 1.3} \times 100\% = \mathbf{35\%\ MgCO_3}$$

5.131 $\frac{0.0200 \text{ mol NaOH}}{\text{L}} \times 0.0152 \text{ L} \times \frac{1 \text{ mol acid}}{2 \text{ mol NaOH}} \times \frac{176.1 \text{ g acid}}{\text{mol}} \Leftrightarrow 0.0268 \text{ g acid}$

$$\frac{0.0268 \text{ g}}{0.1000 \text{ g sample}} \times 100\% = \mathbf{26.8\%}$$

5.133 (a) $\mathbf{MnO_4^- + 5Fe^{2+} + 8H^+ \rightarrow Mn^{2+} + 5Fe^{3+} + 4H_2O}$

(b) $\frac{0.0281 \text{ mol KMnO}_4}{\text{L}} \times 0.03942 \text{ L} \times \frac{1 \text{ mol MnO}_4^-}{\text{mol KMnO}_4} \times \frac{5 \text{ mol Fe}^{2+}}{\text{mol MnO}_4^-} \times$

$$\frac{1 \text{ mol Fe}_3\text{O}_4}{3 \text{ mol Fe}^{2+}} \times \frac{231.5 \text{ g Fe}_3\text{O}_4}{\text{mol Fe}_3\text{O}_4} \Leftrightarrow 0.427 \text{ g Fe}_3\text{O}_4$$

$$\frac{0.427 \text{ g}}{1.362 \text{ g sample}} \times 100\% = \mathbf{31.4\% \text{ of } Fe_3O_4 \text{ by mass}}$$

6 ENERGY AND ENERGY CHANGES: THERMOCHEMISTRY

6.1 Energy is usually defined as the capacity to do work. Matter can have energy as kinetic energy and as potential energy.

6.3 If the potential energy of an object decreases as it moves away from another object, repulsive forces must exist between the two objects.

6.5 Chemical energy is the term used to describe the potential energy that chemicals have because of the attractions and repulsions between their subatomic particles.

6.7 A ball thrown into the air will have maximum kinetic energy at release. Its kinetic energy decreases as the ball rises and slows but its potential energy increases as it reaches greater height. At its maximum height, it stops thus having zero kinetic energy but maximum potential energy. As it falls the process is reversed--potential energy is decreased and kinetic energy is increased.

6.9 The potential energy possessed by atoms and other atomic-sized particles is due to the attractions and repulsions of the electrically charged particles (nuclei and electrons).

6.11 Heat energy is the same as kinetic energy. Heat flows from a hot object to a cool object through the transfer of kinetic energy as fast moving particles collide with slower moving particles and impart some of their energy to the less energetic particles.

6.13 (a) It is exothermic since heat is given up.
(b) The 2Mg + O_2 must have larger potential energy than the product 2MgO in order for the observed exothermic nature of the reaction to take place.
(c) As potential energy is converted to heat, the kinetic energy of the particles is increasing with the increase in temperature.

6.15 The joule is defined as the energy corresponding to the energy possessed by an object with a mass of 2 kg traveling at the velocity of 1 meter per second.
4.184 joules equals 1 calorie. 4.184 kilojoules equals 1 kilocalorie.

6.17 (a) 345 J x (1 cal/4.184 J) = **82.5 cal**
(b) 546 cal x (4.184 J/cal) = **2.28 x 10^3 J**
(c) 234 kJ x (1 kcal/4.184 kJ) = **55.9 kcal**
(d) 1.257 kcal x (4.184 kJ/1 kcal) = **5.259 kJ**

6.19 Knowing that 1 J = 1 kg x m^2/s^2, one would need to convert 145 lb x $(15.3\ mi/hr)^2$ to kg x m^2/s^2 before it can be equated to joules.

$$K.E = \frac{1}{2} \times 145\ lb \times \left(\frac{15.3\ mi}{hr}\right)^2 \times \frac{1\ kg}{2.205\ lb} \times \left(\frac{5280\ ft}{mi}\right)^2 \times \left(\frac{12\ in.}{ft}\right)^2 \times \left(\frac{1\ m}{39.37\ in.}\right)^2$$

$$\times \left(\frac{1\ hr}{60\ min}\right)^2 \times \left(\frac{1\ min}{60\ sec}\right)^2 = 1.54 \times 10^3\ kg\ m^2/s^2 = \mathbf{1.54 \times 10^3\ J}$$

6.21 $$K.E = \frac{1}{2} \times 2.40 \text{ tons} \times \left(\frac{35.0 \text{ mi}}{\text{hr}}\right)^2 \times \frac{2000 \text{ lb}}{\text{ton}} \times \frac{1 \text{ kg}}{2.20 \text{ lb}} \times \left(\frac{5280 \text{ ft}}{\text{mi}}\right)^2 \times \left(\frac{12 \text{ in.}}{\text{ft}}\right)^2$$

$$\times \left(\frac{1 \text{ m}}{39.37 \text{ in.}}\right)^2 \times \left(\frac{1 \text{ hr}}{3600 \text{ sec}}\right)^2 \times \frac{1 \text{ J}}{1 \text{ kg m}^2 \text{ s}^{-2}} = 2.67 \times 10^5 \text{ J}$$

$$2.67 \times 10^5 \text{ J} \times \frac{\text{g °C}}{4.18 \text{ J}} \times \frac{1}{12.0\text{°C}} = 5.32 \times 10^3 \text{ g or } \mathbf{5.32\ kg}$$

6.23 Force x distance = (515 kg m/s^2) x 40.0 m
= 2.06 x 10^4 kg m^2/s^2
(2.06 x 10^4 kg m^2/sec^2) x (1 J/kg m^2 s^{-2}) = **2.06 x 10^4 J**

6.25 The large specific heat of water is responsible for the moderating effects the oceans have on weather.

6.27 The heat capacity of water per mol of water or its molar heat capacity is:
(18.0 g mol^{-1}) x (4.18 J g^{-1} °C^{-1}) = **75.2 J mol^{-1} °C^{-1}**

6.29 500 g x (4.18 J g^{-1} °C^{-1}) x 24.0°C = **5.02 x 10^4 J**
5.02 x 10^4 J x (1 cal/4.18 J) = **1.20 x 10^4 cal**

6.31 Heat lost by the penny = (Use T to indicate final temperature)
3.14 g x (0.387 J g^{-1} °C^{-1}) x (100°C - T) = ?
Heat gained by the water =
10.0 g x (4.18 J g^{-1} °C^{-1}) x (T - 25.0°C) = ?
The two amounts of heat must be equal; therefore, they can be set equal and the unknown "T" can be solved.

3.14 x 0.387 x (100 - T) = 10.0 x 4.18 x (T - 25.0)
121.5 - 1.215 T = 41.8 T - 1045
43.0 T = 1166.5
T = **27.1°C**

6.33 Heat lost by the metal specimen =
25.467 g x specific heat x (100.0°C - 31.2°C) = ?
Heat gained by the water =
15.0 g x 4.18 J g^{-1} $°C^{-1}$ x (31.2°C - 24.3°C) = ?
Set the two equations equal:
25.467 x specific heat x 68.8 = 15.0 x 4.18 x 6.9
specific heat = **0.25 J g^{-1} $°C^{-1}$**

6.35 Heat capacity = 1347 J/(26.135°C - 25.000°C) = **1187 J/°C**

6.37 (a) -(97.1 kJ/°C)(27.282°C - 25.000°C) = -222 kJ or **-2.22 x 10^5 J**
(b) 222 kJ ÷ 1.00 mol = **-222 kJ mol^{-1}**

6.39 (a) Energy change = -(45.06 kJ/°C) x (26.413°C - 25.000°C) x (1000 J/kJ)
= -6.367 x 10^4 J or **6.367 x 10^4 J liberated**

(b) (6.367 x 10^4 J/1.500 g C_7H_8) x (92.15 g C_7H_8/mol C_7H_8)
= **3.911 x 10^6 J mol^{-1} liberated**

6.41 (a) If a system is isothermal, then not only does temperature not change, but average kinetic energy must also be constant since they are directly related.
(b) In an adiabatic system if the potential energy of the chemical substances decreases, the temperature of the system will increase. Therefore, in a reaction that loses potential energy, we will observe a warming (increased kinetic energy) if the total energy is kept constant (adiabatic).

6.43 A state function is a quantity whose value depends only on the current state of a system and not on prior history. Energy change is always final energy content minus initial energy content and it will always be the same value provided that the states of the initial and final values do not change. Therefore, energy change is a state function. Otherwise one could create energy by different paths and break the law of conservation of energy.

6.45 (a) K.E. = 4.18 J g^{-1} °C^{-1} x 18.0 g mol^{-1} x (45°C - 25°C) = **1.5 x 10^3 J mol^{-1}**
(b) K.E. = 4.18 J g^{-1} °C^{-1} x 18.0 g mol^{-1} x (80°C - 25°C) = **4.1 x 10^3 J mol^{-1}**
(c) K.E. = 4.18 J g^{-1} °C^{-1} x 18.0 g mol^{-1} x (45°C - 80°C) = **-2.6 x 10^3 J mol^{-1}**
If one adds (b) and (c), one obtains (4.1 x 10^3 J mol^{-1} - 2.6 x 10^3 J mol^{-1}) or 1.5 x 10^3 J mol^{-1} - the same value as (a). Therefore, the net change of going from 25°C → 80°C → 45°C is the same as going directly from 25°C to 45°C.

6.47 A perpetual motion machine is one that would produce energy during a cyclic process. Such a machine is impossible because the change in energy is independent of path (a state function). Therefore, if no energy is lost to or gained from the surroundings, the net change in energy in going from state A to some other state then back to A must be zero. A perpetual motion machine would violate the law of conservation of energy.

6.49 A substance is in its standard state when it is at a temperature of 25°C and a pressure of 1 atm.

6.51 $SO_2 + 1/2O_2 \rightarrow SO_3$ should not be labeled as a ΔH_f since it is not the formation of SO_3 from its elements.

6.53

$1/2HCHO_2(\ell) + 1/2H_2O(\ell)$	$\rightarrow 1/2CH_3OH(\ell) + 1/2O_2(g)$	205.5 kJ
$1/2CO(g) + H_2(g)$	$\rightarrow 1/2CH_3OH(\ell)$	-64. kJ
$1/2HCHO_2$	$\rightarrow 1/2CO(g) + 1/2H_2O(\ell)$	-16.5 kJ
$HCHO_2(\ell) + H_2(g)$	$\rightarrow CH_3OH(\ell) + 1/2O_2(g)$	$\Delta H°$ = **125 kJ**

6.55

$1/2CaO + 1/2Cl_2$	$\rightarrow 1/2CaOCl_2$	$\Delta H°$ = -55.45 kJ
$1/2H_2O + 1/2CaOCl_2 + NaBr$	$\rightarrow NaCl + 1/2Ca(OH)_2 + 1/2Br_2$	$\Delta H°$ = -30.1 kJ
$1/2Ca(OH)_2$	$\rightarrow 1/2CaO + 1/2H_2O$	$\Delta H°$ = +32.55 kJ
$1/2Cl_2 + NaBr$	$\rightarrow NaCl + 1/2Br_2$	$\Delta H°$ = **-53.0 kJ**

6.57

$12/7NH_3(g) + 3O_2(g) \rightarrow 12/7NO_2(g) + 18/7H_2O(g)$	$\Delta H° = -485.1\ kJ$
$12/7NO_2(g) + 16/7NH_3(g) \rightarrow 2N_2(g) + 24/7H_2O(g)$	$\Delta H° = -782.9\ kJ$
$4NH_3(g) + 3O_2(g) \rightarrow 2N_2(g) + 6H_2O(g)$	$\mathbf{\Delta H° = -1268\ kJ}$

6.59

$6Mg(s) + 3O_2(g) \rightarrow 6MgO(s)$	$\Delta H° = -3(1203\ kJ)$
$3Mg(s) + N_2(g) \rightarrow Mg_3N_2(s)$	$\Delta H° = -1(463\ kJ)$
$6MgO(s) + Mg_3N_2(s) \rightarrow 8Mg(s) + Mg(NO_3)_2(s)$	$\Delta H° = -1(-3884\ kJ)$
$Mg(s) + N_2(g) + 3O_2(g) \rightarrow Mg(NO_3)_2(s)$	$\mathbf{\Delta H° = -188\ kJ}$

6.61

Reaction	$\Delta H°$
(1) $2N_2(g) + 5O_2(g) \rightarrow 2N_2O_5(g)$	?
(2) $2N_2(g) + 6O_2(g) + 2H_2(g) \rightarrow 4HNO_3(\ell)$	4 (-174 kJ)
(3) $4HNO_3(\ell) \rightarrow 2N_2O_5(g) + 2H_2O(\ell)$	- 2 (-76.6 kJ)
(4) $2H_2O(\ell) \rightarrow 2H_2(g) + O_2(g)$	-1 (-571.5 kJ)

(2) + (3) + (4) = (1) $\quad \Delta H° = 4(-174) + [-2(-76.6)] + [-1(-571.5)]$

$\mathbf{\Delta H° = 29\ kJ}$

6.63 $H_2O(\ell) \rightarrow H_2O(g) \quad \Delta H°\ (reaction) = \Delta H°\ (evaporation)$

$\Delta H_{vap} = (-242) - (-286) = +44\ kJ\ mol^{-1}$

$(44\ kJ\ mol^{-1})\ (10.0\ g\ H_2O)\ (1\ mol/18.02\ g) =$ **24 kJ**

6.65 $\Delta H° = (sum\ \Delta H_f°\ products) - (sum\ \Delta H_f°\ reactants)$

(a) $\Delta H° = [(-1676) + 2(0)] - [2(0) + (-822.2)] =$ **-854 kJ**
(b) $\Delta H° = [(-910.0 + 2(-242)] - [(+33) + 0] =$ **-1427 kJ**
(c) $\Delta H° = [-1433] - [(-635.5) + (-396)] =$ **-402 kJ**
(d) $\Delta H° = [(0) + (-242)] - [(-155) + 0] =$ **-87 kJ**
(e) $\Delta H° = [-84.5] - [(+51.9) + 0] =$ **-136.4 kJ**

6.67 $\Delta H = 4.18\ J\ g^{-1}\ °C^{-1} \times 350\ mL \times 1\ g\ mL^{-1} \times (30.00°C - 25.00°C)$

$= 7315\ J$ or $7.315\ kJ$ for the reaction of $0.150\ L \times 1.00\ mol\ L^{-1}\ HCl$

(HCl is the limiting reactant)

$\Delta H = 7.315\ kJ\ /0.150\ mol =$ **48.8 kJ /mol H^+**

6.69 Using the value for heat of combustion one can construct the following:

$$\frac{1\ mol\ C_6H_{12}O_6}{2820\ kJ\ (total)} \times \frac{100\ kJ\ (total)}{60\ kJ\ (available\ as\ heat)} \times \frac{5900\ kJ\ (as\ heat)}{hr}$$

$$\times \frac{180.2\ g\ C_6H_{12}O_6}{mol\ C_6H_{12}O_6} = 628.4\ g\ C_6H_{12}O_6\ hr^{-1}$$

or to two significant figures **630 g hr^{-1}**

6.71 $\Delta H° = (\text{sum } \Delta H_f°\ \text{products}) - (\text{sum } \Delta H_f°\ \text{reactants})$

$FeO(s) + CO(g) \rightarrow Fe(s) + CO_2(g) \qquad \Delta H = -17kJ$

$\Delta H°\ (\text{reaction}) = [\Delta H_f°\ _{Fe} + \Delta H_f°\ _{CO_2}] - [\Delta H_f°\ _{FeO} + \Delta H_f°\ _{CO}]$

$-17\ kJ = [0 + (-394)] - [\Delta H_f°\ _{FeO} + (-110)]$

$\Delta H_f°$ (FeO(s)) $= -267\ kJ\ /mol$

7 ELECTRONIC STRUCTURE AND THE PERIODIC TABLE

7.1 See Figure 7.1. Wavelength, λ, is the distance between consecutive peaks or troughs in a wave. Frequency, ν, is the number of peaks passing a given point per second. They are related to each other by the equation $\lambda \cdot \nu = c$ (equation 7.1) where c is the speed of light.

7.3 SI unit of frequency is the hertz, 1 Hz = 1 s^{-1}. Units for wavelengths are chosen so that the numbers are simple to comprehend. Thus, 320 nm is easier to comprehend than 3.20×10^{-7} m. The visible region of the spectrum runs from about 400 nm to 700 nm.

7.5 (Shortest wavelengths to longest wavelengths) gamma rays, x rays, ultraviolet light, visible light, infrared light, microwaves, TV waves

7.7 Wavelength x frequency = speed of light
(a) wavelength x (8.0×10^{15} s^{-1}) = (3.0×10^{8} m s^{-1})
wavelength = 3.75×10^{-8} m or 2 significant figures: 3.8×10^{-8} m or **38 nm**

(b) 200.0 nm (10^{-9} m/nm) x frequency = speed of light in m s^{-1}
(200.0×10^{-9} m) x ν = (3.00×10^{8} m s^{-1})
$\nu = 1.50 \times 10^{15}$ s^{-1} or **1.50×10^{15} Hz** (Only 3 significant figures because the speed of light is not an exact number.)

7.9 (a) **FM** 101.1 MHz = 101.1×10^{6} Hz = 101.1×10^{6} s^{-1}
$\lambda \times \nu = c$ $\lambda \times (101.1 \times 10^{6}\ s^{-1}) = (3.00 \times 10^{8}\ m\ s^{-1})$ **λ = 2.97 m** (3 significant figures)

(b) **AM** 880 kHz = 880×10^{3} Hz = 880×10^{3} s^{-1}
$\lambda \times (880 \times 10^{3}\ s^{-1}) = 3.00 \times 10^{8}\ m\ s^{-1}$ **λ = 341 m**

7.11 A line spectrum results when the light emitted does not contain radiation of all wavelengths as is needed for a continuous spectrum. A continuous spectrum can be obtained by passing sunlight through a prism. If light emitted by a gas discharge tube is passed through a prism, a line spectrum is obtained.

7.13 The line spectra is also known as the atomic emission spectra, emission spectra and atomic spectra.

7.15 The visible lines of the atomic spectrum of hydrogen are at 410.3 nm, 432.4 nm, 486.3 nm, and 656.4 nm. The frequency of each of these is:
(a) (for 410.3 nm) $\lambda \times \nu = c$: (410.3×10^{-9} m) x ν = 3.00×10^{8} m s^{-1}
ν = **7.31×10^{14} Hz**
(b) (for 432.4 nm) ν = **6.94×10^{14} Hz**
(c) (for 486.3 nm) ν = **6.17×10^{14} Hz**
(d) (for 656.4 nm) ν = **4.57×10^{14} Hz**

7.17 Each element has its own characteristic set of emission spectrum lines which can be used to identify the presence of an element in the presence of other elements.

7.19 A photon is a tiny packet, or quanta, of light energy.

7.21 (a) $E = h\nu = (6.63 \times 10^{-34}\ J\ s)(3 \times 10^{15}\ s^{-1}) =$ **2×10^{-18} J**

(b) $E = hc/\lambda = 2 \times 10^{-20}\ J$

$(6.63 \times 10^{-34}\ J\ s)\ (3.00 \times 10^{8}\ m\ s^{-1})/\lambda = 2 \times 10^{-20}\ J$

Solving for λ gives: **$\lambda = 1 \times 10^{-5}$ m**

7.23 (a) $E = h\nu = (6.63 \times 10^{-34}\ J\ s/photon)(2.6 \times 10^{14}\ s^{-1}) = 1.7 \times 10^{-19}$ J/photon

$(1.7 \times 10^{-19}$ J/photon) x $(6.02 \times 10^{23}$ photons/mol) = **1.0×10^{5} J /mol**

= 100 kJ /mol

(b) $E = hc/\lambda = (6.63 \times 10^{-34}\ J\ s/photon)(3.00 \times 10^{8}\ m\ s^{-1})/(546 \times 10^{-9}\ m)$

$= 3.64 \times 10^{-19}$ J/photon

$(3.64 \times 10^{-19}$ J/photon) x $(6.02 \times 10^{23}$ photons/mol) = **2.19×10^{5} J/mol**

= 219 kJ /mol

7.25 The Bohr model imagined that the electrons travel around the nucleus in orbits of fixed size and energy. For this model an equation could be mathematically derived for the wavelengths of the light emitted by hydrogen when it produced its atomic spectrum. The model (theory) failed to correctly calculate energies for any atoms more complex than hydrogen.

7.27 (a) $$\frac{1}{\lambda} = 109{,}678\ cm^{-1}\left[\frac{1}{2^2} - \frac{1}{4^2}\right]$$

$\lambda = 4.86273 \times 10^{-5}\ cm =$

$\lambda = 4.86273 \times 10^{-5}\ cm \times 10^{-2}\ m/cm \times 1\ nm/10^{-9}\ m =$ **486.273 nm**

(b) $$\frac{1}{\lambda} = 109{,}678\ cm^{-1}\left[\frac{1}{3^2} - \frac{1}{6^2}\right]$$ $\lambda =$ **1.09411×10^{-4} cm = 1094.11 nm**

7.29 (a) $E = -A/n^2 = -2.18 \times 10^{-18}\ J/(3)^2 = \mathbf{-2.42 \times 10^{-19}\ J}$
(b) $E = -A/n^2 = -2.18 \times 10^{-18}\ J/(2)^2 = \mathbf{-5.45 \times 10^{-19}\ J}$
(c) $\Delta E = (-2.42 \times 10^{-19}\ J) - (-5.45 \times 10^{-19}\ J) = \mathbf{3.03 \times 10^{-19}\ J}$
(d) $\nu = E/h = 3.03 \times 10^{-19}\ J/6.63 \times 10^{-34}\ J\ s = \mathbf{4.57 \times 10^{14}\ Hz}$
(e) $\lambda = c/\nu = (3.00 \times 10^8\ m\ s^{-1}/4.57 \times 10^{14}\ s^{-1}) \times 1\ nm/10^{-9}\ m = \mathbf{656\ nm}$
(f) 656 nm is between the orange and red regions of the visible spectrum (Figure 7.2) Therefore, one would guess that the color of light having a wavelength of 656 nm would be **red-orange or red.**

7.31 $\lambda = \dfrac{h}{m\nu}$ $\quad \nu = c = 3.00 \times 10^8\ m\ s^{-1}$; $\lambda = 589\ nm = 589 \times 10^{-9}\ m$

$$h = 6.63 \times 10^{-34}\ J\ s = 6.63 \times 10^{-34}\ kg\ m^2/s$$

$$m = \frac{6.63 \times 10^{-34}\ kg\ m^2/s}{(589 \times 10^{-9}\ m)(3.00 \times 10^8\ m\ s^{-1})} = 3.75 \times 10^{-36}\ kg\ \text{(per photon)}$$

$$m = \left(\frac{3.75 \times 10^{-36}\ kg}{photon}\right) \times \left(\frac{6.02 \times 10^{23}\ photon}{mole}\right) \times \left(\frac{1000\ g}{kg}\right) = \mathbf{2.26 \times 10^{-9}\ g}$$

7.33 Because the wavelengths are too short to be detected.

7.35 Electrons and other subatomic particles can be used to produce diffraction patterns. Diffraction patterns of electrons can only be explained as the result of the wave properties of the electrons.

7.37 The principle quantum number symbol is **n**. Allowed values of n are: **1, 2, 3, 4, etc.** The larger the value of n, the greater the average energy of the levels belonging to its associated shell and the larger the size of the wave function.

7.39 The magnetic quantum number symbol is $\mathbf{m_\ell}$. Allowed values of m_ℓ are: integer values that range from **$-\ell$ to $+\ell$ including zero.** The magnetic quantum number serves to determine an orbital's orientation in space relative to the other orbitals.

7.41 **4**

7.43 **f**

7.45 The ground state is the state of lowest energy.

7.47 In order of increasing energy they are: s < p < d < f

7.49 The spin of the electron causes the electron to act as a tiny electromagnet.

7.51 The Pauli exclusion principle states that no two electrons in any one atom can have all four quantum numbers the same. This limits the number of electrons in any given orbital to two -- one with a positive spin, the other with a negative spin.

7.53 **18**

7.55 Unpaired electrons give atoms, molecules or ions that are paramagnetic. Paramagnetic species are weakly attracted to a magnetic field. Atoms, molecules or ions that have no unpaired electrons are slightly repelled by a magnetic field and are said to be diamagnetic.

7.57 Predicted electron configurations based on position in periodic table.

P; $1s^2\ 2s^2\ 2p^6\ 3s^2\ 3p^3$
Ni; $1s^2\ 2s^2\ 2p^6\ 3s^2\ 3p^6\ 3d^8\ 4s^2$
As; $1s^2\ 2s^2\ 2p^6\ 3s^2\ 3p^6\ 3d^{10}\ 4s^2\ 4p^3$
Ba; $1s^2\ 2s^2\ 2p^6\ 3s^2\ 3p^6\ 3d^{10}\ 4s^2\ 4p^6\ 4d^{10}\ 5s^2\ 5p^6\ 6s^2$
Rh; $1s^2\ 2s^2\ 2p^6\ 3s^2\ 3p^6\ 3d^{10}\ 4s^2\ 4p^6\ 4d^7\ 5s^2$ expected ($4d^8\ 5s^1$ actual)
Ho; $1s^2\ 2s^2\ 2p^6\ 3s^2\ 3p^6\ 3d^{10}\ 4s^2\ 4p^6\ 4d^{10}\ 4f^{11}\ 5s^2\ 5p^6\ 5d^0\ 6s^2$
Ge; $1s^2\ 2s^2\ 2p^6\ 3s^2\ 3p^6\ 3d^{10}\ 4s^2\ 4p^2$

7.59 K $4s^1$; Al $3s^2 3p^1$; F $2s^2 2p^5$; S $3s^2 3p^4$; Tl $6s^2 6p^1$; Bi $6s^2 6p^3$

7.61 Since there are no d or f orbitals in periods 1 and 2 and the 3d subshell does not begin to fill until the 3p and 4s subshells have been filled, there are no orbitals to fill between the 3s orbital of Mg and the 3p orbital of Al.

7.63 (a) P

1s	2s	2p	3s	3p
↑↓	↑↓	↑↓ ↑↓ ↑↓	↑↓	↑ ↑ ↑

(b) Ca

1s	2s	2p	3s	3p	4s
↑↓	↑↓	↑↓ ↑↓ ↑↓	↑↓	↑↓ ↑↓ ↑↓	↑↓

7.65 Cd, Sr and Kr are diamagnetic since they have no unpaired electrons.

7.67 Yes, the value of ℓ does reflect the shape of an orbital. When $\ell = 0$, we are describing a spherical s orbital. When $\ell = 1$, the double lobed p orbital, etc. Greater values of n represent larger orbitals.

7.69 The size increases from 1s to 2s, etc., and more nodes occur. Their overall shape, however, is spherical.

7.71 See Figure 7.18

7.73 An atom or ion has no fixed outer limits. Atomic and ionic sizes are usually given in angstroms, nanometers, or picometers.

7.75 Effective nuclear charge refers to the residual net charge felt by the outer valence electrons.

7.77 (a) Be, 2+ (b) Mg, 2+ (c) Ca, 2+ (d) Sr, 2+ (e) Ba, 2+

7.79 (a) Ca^{2+} [Ar] (b) S^{2-} [Ar] (c) Cl^{-} [Ar] (d) K^{+} [Ar]
These are all $1s^2\ 2s^2\ 2p^6\ 3s^2\ 3p^6$

7.81 Size trends within the periodic table would predict that Sn is largest.

7.83 The ions N^{3-}, O^{2-}, and F^{-} are isoelectronic (i.e., they have identical electron configurations). The effective nuclear charge is increasing N < O < F. This leads to greater attraction for the outer-shell electrons which are pulled closer to the nucleus decreasing the size of the ion. The electron-electron repulsions are the same so the variation of effective nuclear charge is primarily responsible for their size variation.

7.85 When there are fewer electrons, the interelectron repulsions are less and the outer shell can contract in size under the influence of the nuclear charge.

7.87 As we move from left to right across a period, the increased effective nuclear charge causes the shell to shrink in size and also makes it more difficult to remove an electron. Therefore, ionization energy increases from left to right across the periodic table with irregularities due to filled and half-filled subshells.

7.89 (a) Cl (b) S (c) P (predicted from general trends; actually, this is an exception; the EA for As is more exothermic.) (d) S

7.91 $$\frac{1}{\lambda} = 109{,}678\ \text{cm}^{-1}\left(\frac{1}{1^2} - \frac{1}{\infty^2}\right) = 109{,}678\ \text{cm}^{-1}$$

$$E = h\nu = \frac{hc}{\lambda} = \left(6.63 \times 10^{-34}\ \text{J s}\right)\left(3.00 \times 10^{8}\ \text{m s}^{-1}\right)\left(109{,}678\ \text{cm}^{-1}\right)\left(\frac{100\ \text{cm}}{1\ \text{m}}\right)$$

$$= 2.18 \times 10^{-18}\ \text{J}$$

$$\left(2.18 \times 10^{-18}\ \frac{\text{J}}{\text{atom}}\right)\left(6.02 \times 10^{23}\ \frac{\text{atom}}{\text{mol}}\right)\left(\frac{1\ \text{kJ}}{1000\ \text{J}}\right) = 1310\ \text{kJ/mol}$$

vs 1312 kJ/mol in Table 7.4

7.93 No. There is no way that all electrons can be paired if there is an odd number of electrons. Therefore, if there is an odd number of electrons, the atom cannot be diamagnetic.

7.95 Elements at the center of period 6 are so dense due to the **lanthanide contraction.**

7.97 Z = 114 would belong to group **IVA**. Its configuration would be:

[Rn] $5f^{14}\ 6d^{10}\ 7s^2\ 7p^2$

7.99 Each graph will show a large jump in ionization energy as one goes beyond the removal of valence electrons. Until all valence electrons are removed, the increase in ionization energy is a gradually increasing phenomenon.

8 CHEMICAL BONDING:

GENERAL CONCEPTS

8.1 An <u>ionic bond</u> results from the attraction between oppositely charged ions.

8.3 (a) $Ba^{2+} \Leftrightarrow [Xe]$ (b) $Se^{2-} \Leftrightarrow [Kr]$ (c) $Al^{3+} \Leftrightarrow [Ne]$
(d) $Na^{+} \Leftrightarrow [Ne]$ (e) $Br^{-} \Leftrightarrow [Kr]$

8.5 $AlCl_4$ does not form because aluminum does not form the Al^{4+} ion under normal circumstances. Likewise, Na_3O does not form because oxygen does not form a three minus ion.

8.7 (a)

$Na(g) \rightarrow Na^{+}(g) + e^{-}$	495.8 kJ
$Cl(g) + e^{-} \rightarrow Cl^{-}(g)$	-348 kJ
$Na(g) + Cl(g) \rightarrow Na^{+}(g) + Cl^{-}(g)$	**148 kJ**

(continued)

8.7 (continued)

(b) $Na(g) \rightarrow Na^+(g) + e^-$ 496 kJ

$Na^+(g) \rightarrow Na^{2+}(g) + e^-$ 4565 kJ

$2Cl(g) + 2e^- \rightarrow 2Cl^-(g)$ 2(-348 kJ)

$Na(g) + 2Cl(g) \rightarrow Na^{2+}(g) + 2Cl^-(g)$ **4,365 kJ**

The lattice energy of $NaCl_2$ would have to be more than 29.5 times the lattice energy of NaCl before $NaCl_2$ would become more stable than NaCl.

8.9 In Li^+ $[:\ddot{\underset{..}{F}}:]^-$ the **fluoride obeys** the octet rule; **lithium does not.**

8.11 KF(s) is more stable than K(s) and $F_2(g)$ because of the large lattice energy value.

8.13 (a) Zn^{2+} [Ar] $3d^{10}$ (b) Sn^{2+} [Kr] $4d^{10}$ $5s^2$ (c) Bi^{3+} [Xe] $4f^{14}$ $5d^{10}$ $6s^2$
(d) Cr^{2+} [Ar] $3d^4$ (e) Fe^{3+} [Ar] $3d^5$ (f) Ag^+ [Kr] $4d^{10}$

8.15 (a) both are [Ar]
(d) both are [Kr]
(e) both are [Ne]

8.17 For many of them, their outer shell has two electrons. Loss of the two electrons produces a pseudonoble gas configuration.

8.19 (a) $:\dot{\underset{..}{Se}}\cdot$ (b) $:\ddot{\underset{..}{Br}}\cdot$ (c) $\cdot\dot{Al}\cdot$ (d) $\cdot$**Ba**$\cdot$ (e) $\cdot\dot{\underset{\cdot}{Ge}}\cdot$ (f) $\cdot\ddot{\underset{\cdot}{P}}\cdot$

8.21 $Mg: + Mg: + \cdot\dot{\underset{\cdot}{C}}\cdot \rightarrow 2Mg^{2+}, [:\ddot{\underset{..}{C}}:]^{4-}$ or Mg_2C

8.23 (a) **Two** (b) $\cdot\dot{\underset{\cdot}{C}}\cdot$ **Four** (c) Lewis symbols are written in a way that reflects the number of unpaired electrons that are typically involved in bonding and not necessarily the number of unpaired electrons on the atom in its ground state.

8.25 Neither Cl nor F can form a stable cation. For both Cl and F the first ionization energy is much greater than the electron affinity. Both strive to achieve an octet by gaining an electron, not by losing electrons. Therefore, no energetically favored combination of cation and anion can be formed from Cl and F.

8.27 (a) $:\dot{\underset{\cdot}{N}}\cdot + 3H\cdot \rightarrow :\overset{H}{\underset{H}{\ddot{\underset{\cdot\cdot}{N}}}}:H$ (b) $2H\cdot + \cdot\dot{\underset{\cdot\cdot}{O}}: \rightarrow H:\overset{H}{\ddot{\underset{\cdot\cdot}{O}}}:$ (c) $H\cdot + \cdot\ddot{\underset{\cdot\cdot}{F}}: \rightarrow H:\ddot{\underset{\cdot\cdot}{F}}:$

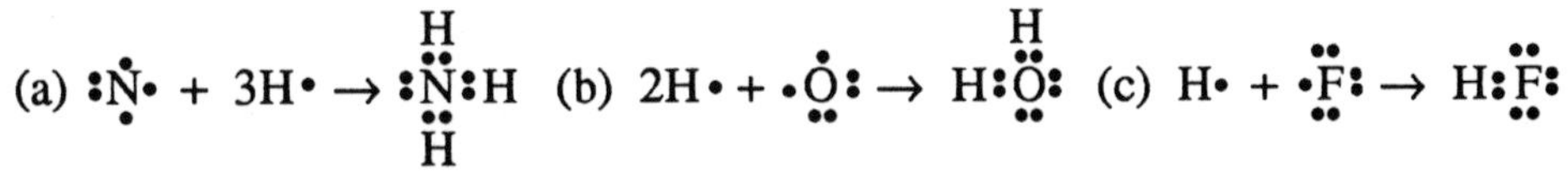

8.29 There are no unpaired electrons in any of these compounds. NH_3, H_2O, and HCl each have unshared but not unpaired electrons. (See Question 8.27).

8.31 (a) $:\ddot{\underset{\cdot\cdot}{Cl}}:Be:\ddot{\underset{\cdot\cdot}{Cl}}:$ and $:\ddot{\underset{\cdot\cdot}{Cl}}:\overset{:\ddot{Cl}:}{\ddot{B}}:\ddot{\underset{\cdot\cdot}{Cl}}:$ (b) **4** in $BeCl_2$ and **6** in BCl_3

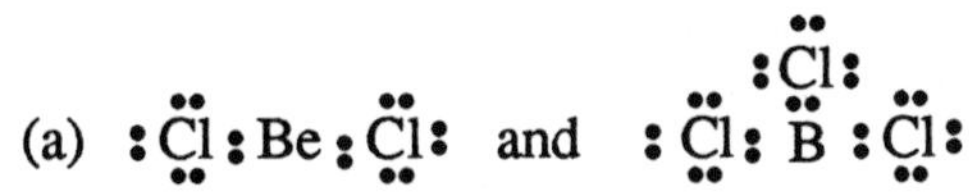

8.33 (a) **4** (b) **4** (c) **9** (d) **9**

8.35 (a) H—N—H (with H bonded below N) (b) Cl—P—Cl (with Cl bonded below P) (c) Cl—S—Cl (bent)

(d) $[O{=}N{-}O]^-$ (e) BrF_5: F—Br—F, with F above and two F below Br (f) $[Cl—P—Cl]^+$, with Cl above and Cl below P

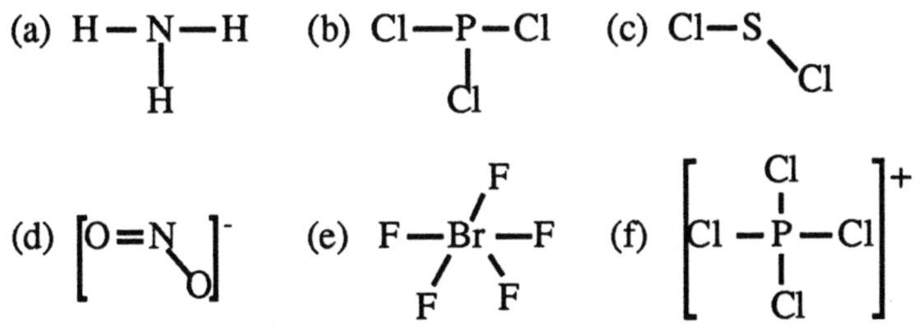

8.37 (a) **8** (b) **26** (c) **20** (d) **18** (e) **42** (f) **32**

8.39 The maximum number of covalent bonds formed by a hydrogen atom is one.

8.41 :Cl: Cl: O / :S::O   F / :O: F: H / H: Sn: H / H   H C::C H / H H   :Cl: / :S :Cl:

8.43 [O / N : O / :O]$^-$ [:N::: O:]$^+$ [:O: / N :: O:]$^-$ [:O: / C :: O: / O]$^{2-}$

8.45 ClF_3, SF_4, IF_7, NO_2, BCl_3

8.47 (a) Each C-Cl bond has a bond order of 1.

(b) In HCN the H-C bond order is 1 and the C N bond order is 3.

(c) In CO_2 each C=O has a bond order of 2.

(d) In NO^+ the bond order is 3.

(e) In CH_3NCO the bond order between each hydrogen and the carbon is 1, between one of the carbons and nitrogen it is 1, between nitrogen and the other carbon it is 2 and between carbon and the oxygen it is 2.

8.49 Bond length decreases as bond order increases because the additional electron density between the nuclei causes the nuclei to be pulled together.

8.51 The greater the bond order the shorter and stronger the bond and the greater the vibrational frequency. Strong bonds will vibrate faster with shorter amplitude just as will a strong spring compared to a weak spring. Infrared absorption spectra are used to measure bond vibrational frequencies. When infrared radiation interacts with a substance, the infrared frequencies that are the same as the vibrational frequencies of the bonds in the substance are absorbed.

8.53 A resonance hybrid is the true structure of a compound. It is a composite of the contributing structures that one can draw. We use resonance because it is impossible to draw a single electron-dot formula that obeys the octet rule and is consistent with experimental facts at the same time.

8.55 H−O−N(=O)−O ⟷ H−O−N(−O)=O

O=Se−O ⟷ O−Se=O

(O=)Se(−O)(−O) ⟷ (O−)Se(=O)(−O) ⟷ (O−)Se(−O)(=O)

(continued)

8.55 (continued)

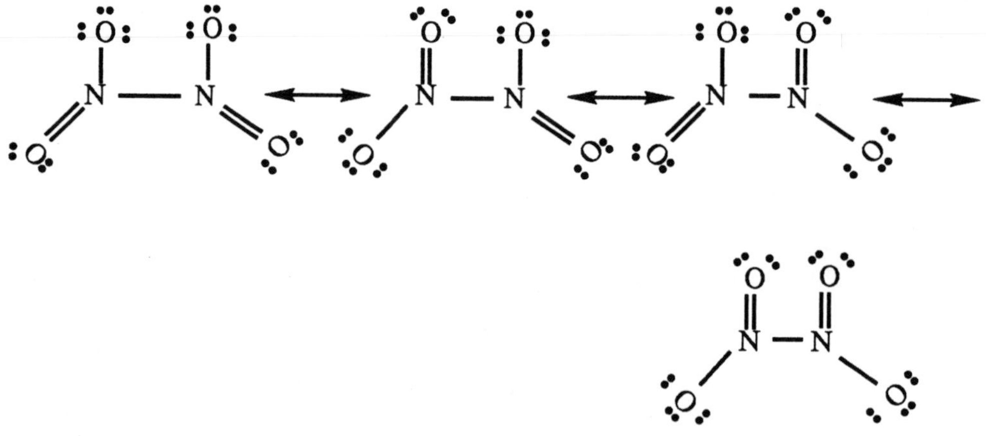

8.57 The N-O bond order in NO_2^- is an average of 1.5 and in NO_3^- it is an average of 1.3. Therefore, the **bond length** should be a little **shorter** and the **bond energy** a little **higher** in NO_2^-.

8.59

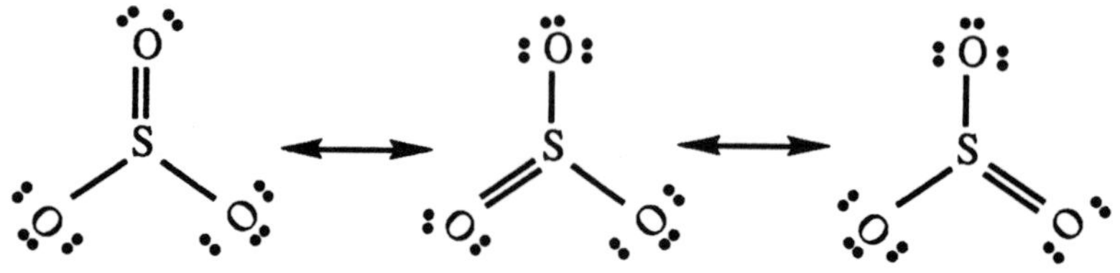

Molecule	Average Bond Order	Bond Length	Bond Energy	Vib.Frequency
SO_2	1.5	shorter	more	higher
SO_3	1.3	longer	less	lower

8.61 (See question 8.59)

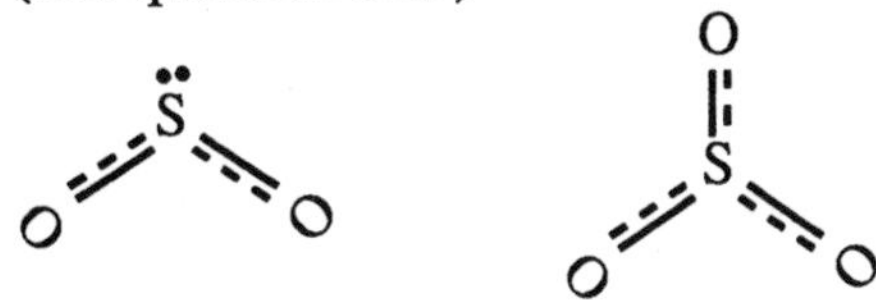

8.63 The lowering of the energy of the molecule compared to the energy of any one of its contributing resonance structures is called resonance energy.

8.65

$$\left[\; :\ddot{O}- \overset{:O:\;\;}{\overset{\|}{S}} = \underset{\cdot\cdot}{S}\,: \right]^{2-} \quad \text{(bottom: } -\,:\ddot{O}:\text{ single-bonded to center S)}$$

	Formal Charge
center S	6 - 6 = 0
top O	6 - 2 - 4 = 0
right S	6 - 2 - 4 = 0
left and bottom O's	2(6 - 1 -6 =-1)
Total	**- 2**

8.67 Answers to this question should also include Lewis dot structures.

(a) H = 1 - 1 = **0**
Cl = 7 - 4 = **+3**
O (between Cl and H) = 6 - 2 - 4 = **0**
O (other 3 oxygens) = 6 - 1 - 6 = **-1**

(b) H = 1 - 1 = **0**
Cl = 7 - 2- 4 = **+1**
O (between Cl and H) = 6 - 2 - 4 = **0**
O (other oxygen) = 6 - 1 - 6 = **-1**

(c) H = 1 - 1 = **0**
S = 6 - 3 - 2 = **+1**
O (2 between H's and the S) = 6 - 2 - 4 = **0**
O (not bonded to H) = 6 - 1 - 6 = **-1**

(d) H = 1 - 1 = **0**
P = 5 - 4 = **+1**
O (3 between H's and the P) = 6 - 2 - 4 = **0**
O (not bonded to H) = 6 - 1 - 6 = **-1**

8.69 In each of the molecules in Question 8.67 a better Lewis structure can be drawn if the octet rule is not obeyed.

(a) 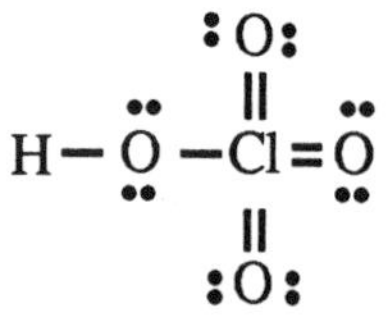 H—O—Cl(=O)(=O)=O

no formal charges other than zero
no resonance structures

(b) 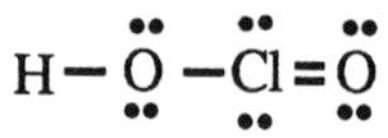 H—O—Cl=O

no formal charges other than zero
no resonance structures

(c) 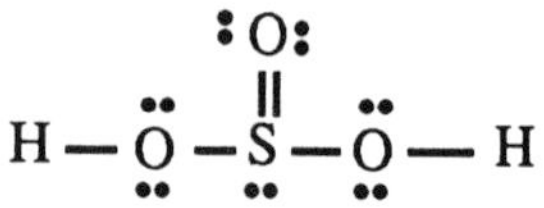 H—O—S(=O)—O—H

no formal charges other than zero
no resonance structures

(d) 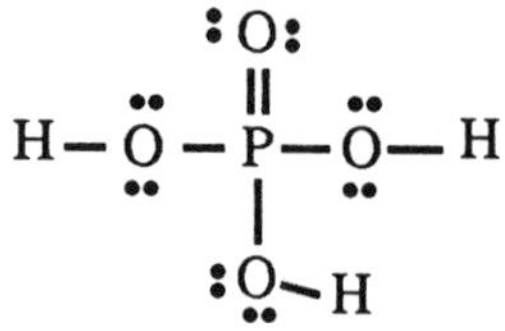 H—O—P(=O)(—O—H)—O—H

no formal charges other than zero
no resonance structures

8.71 $[:N\equiv N^{\oplus}-\ddot{N}:^{(-2)}]^- \longleftrightarrow [:\ddot{N}^{\ominus}=N^{\oplus}=\ddot{N}:^{\ominus}]^- \longleftrightarrow [:\ddot{N}:^{(-2)}-N^{\oplus}\equiv N:]^-$

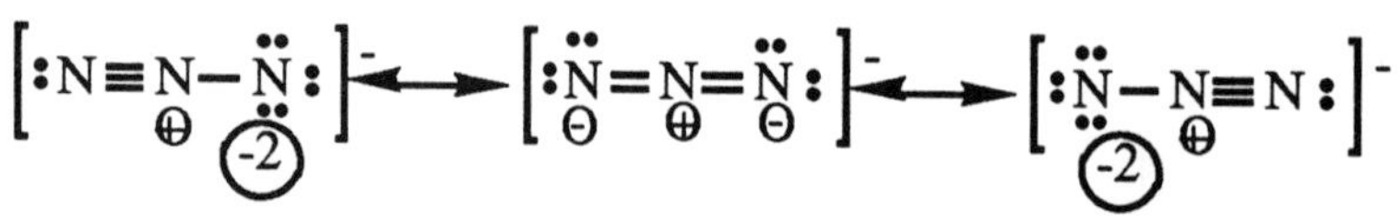

The most stable structure would be the second structure.

8.73 The structure given in this question is rejected due to its having larger formal charges than that structure given in Question 8.70. (Also, since oxygen is one of the most electronegative elements, oxygen does not assume a positive formal charge except when bonded to F. The structure in Question 8.73 has a positive formal charge on the oxygen between the H and the N).

8.75 (a) Average bond order = 5/3 = 1.67

(b) Average bond order = 3/2 =1.5

(c) Average bond order = 5/4 = 1.25

(d) Average bond order = 4/3 = 1.33

(a) Average bond order = 7/4 = 1.75

8.77

:O:
||
:Cl – S — Cl :
||
:O:

All formal charges = zero

8.79

:Cl :
|
:Cl—Al
|
:Cl : + [: Cl:]$^-$ ⟶ [:Cl: | :Cl—Al←Cl : | :Cl :]$^-$

8.81

: F :
|
:F —B
|
: F : + [: F :]$^-$ ⟶ [: F : | : F —B←F : | : F :]$^-$

8.83 A polar molecule is one that has its positive and negative charges separated by a distance. The resulting molecule is said to be a dipole. The dipole moment is the product of the charge on either end of the dipole times the distance between the charges.

8.85 (a) P—F (b) Al—Cl (c) Se—Cl

8.87 NH_3, BCl_3, $MgCl_2$, BeI_2 and NaH

8.89 Rb

8.91

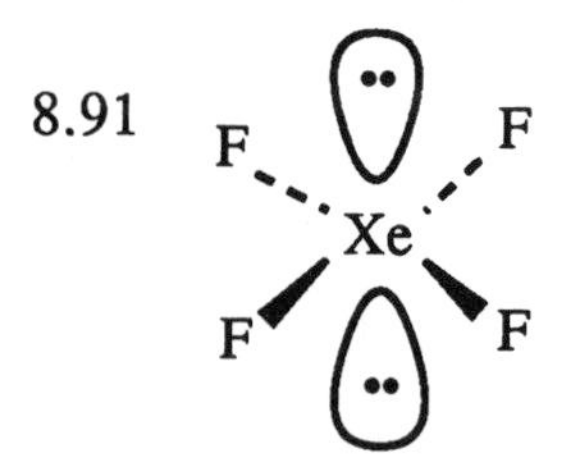

The bonds are polar but, since it is symmetrical, it is a nonpolar molecule.

8.93

Reaction	Energy
1. $Ca(s) \rightarrow Ca(g)$	192 kJ
2. $Ca(g) \rightarrow Ca^{+}(s) + e^{-}$	589.5 kJ
3. $Ca^{+}(g) \rightarrow Ca^{2+}(g) + e^{-}$	1146 kJ
4. $Cl_2(g) \rightarrow 2Cl(g)$	238 kJ
5. $2Cl(g) + 2e^{-} \rightarrow 2Cl^{-}(g)$	2(-348) kJ
6. $Ca^{2+}(g) + 2Cl^{-}(g) \rightarrow CaCl_2(s)$	lattice energy
7. $Ca(s) + Cl_2(g) \rightarrow CaCl_2(s)$	-795 kJ

The sum of reactions 1-6 equals reaction 7. Therefore, 192 kJ + 589.5 kJ + 1146 kJ + 238 kJ + 2(-348) kJ + lattice energy must equal -795 kJ.
Lattice energy = **-2,264 kJ** per mol $CaCl_2$.

8.95 The absolute difference between a calculated bond energy and an experimental bond energy is proportional to the electronegativity difference between the bonded atoms.

Compounds	HF	HCl	HBr	HI
Difference between calc. and exp. bond energies (kJ/mol)	270	94	50	10

Since the electronegativity of hydrogen is constant and less than the electronegativity of the element to which it is bonded, the differences must be due to the presence of the other elements. The change in bond energy indicates a decrease in electronegativity from F to I.

9 COVALENT BONDING AND MOLECULAR STRUCTURE

9.1 See the figures in Section 9.1 for drawing of the linear, planar triangular, tetrahedral, trigonal bipyramidal and octahedral molecular shapes.

9.3 planar triangular, **120°**; tetrahedral, **109.5°**; octahedral, **90°**

9.5 (a) **4** (b) **6** (c) **5**

9.7 (a) MX_3E; trigonal pyramidal (b) MX_4; tetrahedral (c) MX_3; planar triangular

(d) MX_2E; nonlinear or V-shaped (e) MX_4E; unsymmetrical tetrahedral

(f) MX_2E_3; linear (g) MX_5E; square pyramidal (h) MX_4; tetrahedral

(i) MX_6; octahedral (j) MX_5; trigonal bipyramidal

(k) MX_2E; nonlinear, bent, angular or V-shaped

9.9

	(1)	(2)
(a)	planar triangular	planar triangular
(b)	tetrahedral	trigonal pyramidal
(c)	octahedral	octahedral
(d)	tetrahedral	trigonal pyramidal
(e)	tetrahedral	trigonal pyramidal
(f)	tetrahedral	tetrahedral
(g)	planar triangular	planar triangular
(h)	tetrahedral	tetrahedral
(i)	tetrahedral	nonlinear (bent)
(j)	octahedral	square planar

9.11 (a) planar triangular to tetrahedral
(b) trigonal bipyramidal to octahedral
(c) T-shaped to square planar
(d) nonlinear (V-shaped) to unsymmetrical tetrahedral
(e) linear to planar "double" bent

```
H       H
 \     /
  C = C
 /     \
H       H
```

(Each half is planar triangular with the total molecule being coplanar.)
(f) planar triangular to linear

9.13 When two atoms joined by a covalent bond differ in electronegativity, the bond will be polar due to the resulting uneven distribution of electrons. A dipole exists when opposite ends of the molecule carry opposite electrical charges.

9.15 Because SO_2 is V-shaped (bent), its bond dipoles do not cancel each other. The result is a polar molecule. The dipoles in the planar triangular arrangement in SO_3 do cancel and the molecule is nonpolar.

9.17 Planar triangular

9.19 Orbital overlap means that the two orbitals share some common region in space. The pair of electrons associated with a covalent bond is shared between the two atoms in this region and the strength of the covalent bond is proportional to the amount of overlap.

9.21 The s orbital on hydrogen with one electron overlaps with the p orbital with one electron on the Cl atom. See Figure 9.9. Chlorine and fluorine are interchangeable in this type of consideration.

9.23 Hybrid orbitals are produced when two or more atomic orbitals are mixed, producing a new set of orbitals. In hybrid orbitals one lobe is much larger than the other. Some atoms prefer to use hybrid orbitals because hybrid orbitals tend to form stronger covalent bonds (overlap more effectively) than those from ordinary atomic orbitals and hybridization allows the formation of equivalent bonds.

9.25 (a) 109.5° (b) 120° (c) 180° (d) 90°

9.27 We must employ hybrid orbitals in order to account for the H - C - H bond angle of 109.5° in CH_4 and the equivalence of the four C - H bonds.

9.29 Bond angle measurements indicate that the lone pair of electrons project out from the central atom rather than being spread symmetrically as in the s orbital. The reaction of the alkaline properties of NH_3, studied in a later chapter, will also lend experimental evidence to the sp^3 hybridization of nitrogen in NH_3.

9.31 p orbitals

9.33 No. Our explanation is no better than the quality of the experimental evidence about this compound and very similar compounds. There is considerable evidence that indicates that fluorine uses hybrid orbitals.

9.35 (a) planar triangular, sp^2 bonding

(b) tetrahedral, sp^3 bonding

(c) trigonal bipyramidal, sp^3d bonding

(d) octahedral, sp^3d^2 bonding

(e) linear, sp bonding

(f) octahedral, sp^3d^2 bonding

(g) trigonal pyramidal, sp^3 bonding

(h) unsymmetrical tetrahedral, sp^3d bonding

(i) tetrahedral, sp^3 bonding

9.37 (a) sp^2 (b) sp^3 (c) sp^3d^2 (d) sp^3

(e) sp^3 (f) sp^3 (g) sp^2 (h) sp^3

(i) sp^3 (j) sp^3d^2

9.39 Si has a relatively low-energy d subshell in its valence shell, whereas C does not have a d subshell in its valence shell.

9.41 A coordinate covalent bond "exists" in NH_4^+, $AlCl_6^{3-}$, $SbCl_6^-$ and ClO_4^-. After formation, the coordinate covalent bond is identical to the "normal" covalent bond.

9.43 In NH_3, the N - H bonds are polar with the negative ends of the bond dipoles at the nitrogen. The lone pair, if it were in a nonbonded sp^3 orbital, would also produce a contribution to the dipole moment of the molecule, with its negative end pointing <u>away</u> from the nitrogen.

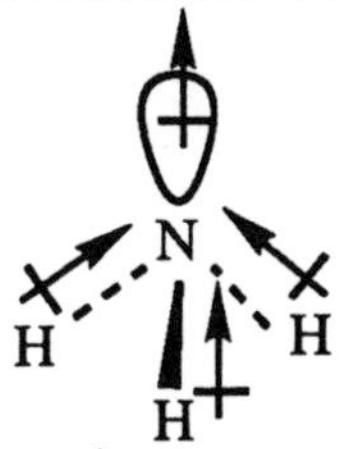

N
H H
H

All dipoles are additive and produce a large net molecular dipole. In NF_3, the fluorines are more electronegative than nitrogen, producing bond dipoles with their positive ends at nitrogen. These, therefore, tend to offset the contribution of the lone pair in the sp^3 orbital on nitrogen, thereby giving a very small net dipole moment for the NF_3 molecule.

N
F F
F

(The bond dipoles tend to cancel the effects of the lone pair dipole. This would not be the case if an s orbital were used for the lone pair and p orbitals were used for the sigma bonds.)

9.45

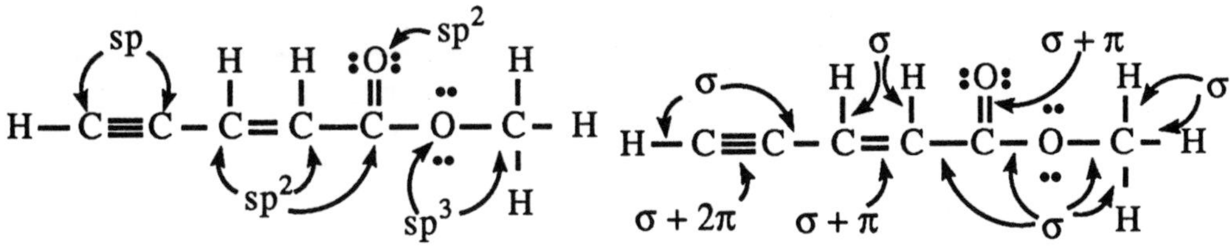

sp, sp^2, sp^2, sp^3

H—C≡C—C=C—C—O—C—H (with H, H, :O:, H, H substituents)

σ, σ + π, σ, σ, σ + 2π, σ + π, σ

H—C≡C—C=C—C—O—C—H

9.47 Valence bond theory shows that the CN^- triple bond consists of one sp-sp σ bond and two p-p π-bonds.

9.49 $[:\ddot{O}=\ddot{N}-\ddot{O}:]^- \longleftrightarrow [:\ddot{O}-\ddot{N}=\ddot{O}:]^-$

sp^2 hybrid orbitals are used for σ bonds

p orbitals are used for π bond

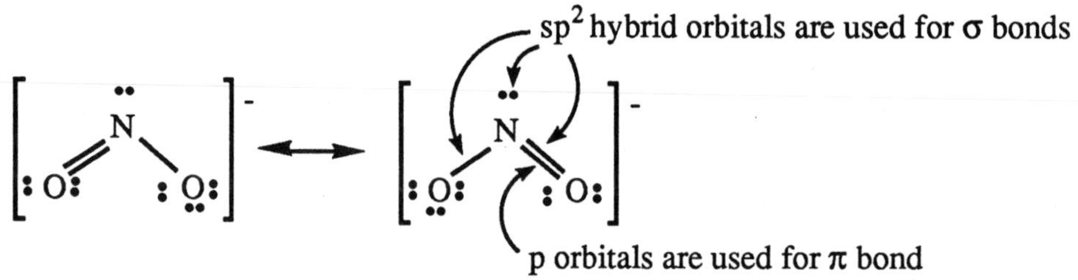

9.51 Because they can exist without involving a pi bond, as they need to form only one bond (a σ bond).

9.53 Diatomic oxygen (O_2) and ozone (O_3).

9.55 White phosphorus consists of P_4 molecules in which each phosphorus atom lies at a corner of a tetrahedron. (See Figure 9.28)

9.57 Compare the structures shown in Figures 9.28 and 9.30.

9.59 Because Si has no apparent tendency to form multiple (π) bonds.

10 CHEMICAL REACTIONS AND THE PERIODIC TABLE

10.1 Metals have positive oxidation states in nearly all of their compounds.

10.3 Since the only effective oxidizing agent in aqueous solutions of HCl is H^+, HCl is called a non-oxidizing acid.

10.5 $2Al(s) + 6HBr(aq) \rightarrow 2AlBr_3(aq) + 3H_2(g)$

10.7 (a) $Ag(s) + NO_3^-(aq) + 2H^+(aq) \rightarrow Ag^+(aq) + NO_2(g) + H_2O$

(b) $3Ag(s) + NO_3^-(aq) + 4H^+(aq) \rightarrow 3Ag^+(aq) + NO(g) + 2H_2O$

10.9 $4Zn(s) + NO_3^-(aq) + 10H^+(aq) \rightarrow 4Zn^{2+}(aq) + NH_4^+(aq) + 3H_2O$

10.11 (a) $Mg(s) + 2HCl(aq) \rightarrow MgCl_2(aq) + H_2(g)$

(b) $2Al(s) + 6HCl(aq) \rightarrow 2AlCl_3(aq) + 3H_2(g)$

10.13 For 10.11: (a) $Mg(s) + 2H^+(aq) \rightarrow Mg^{2+}(aq) + H_2(g)$
(b) $2Al(s) + 6H^+(aq) \rightarrow 2Al^{3+}(aq) + 3H_2(g)$

For 10.12: (a) $2Cr(s) + 6H^+(aq) \rightarrow 2Cr^{3+}(aq) + 3H_2(g)$
(b) $Ni(s) + 2H^+(aq) \rightarrow Ni^{2+}(aq) + H_2(g)$
(The oxidizing agent is H^+ in each of these reactions.)

10.15 In order of increasing ease of oxidation: Ag < Cu < Sn < Cd < Mg

10.17 The easily oxidized metals are located on the left side of the periodic table including IA metals and IIA metals except beryllium. The least easily oxidized metals are located in Period 6 just to the right of center in the block of transition metals.

10.19 They all react with water to liberate hydrogen.

10.21 Ca should react more rapidly because it has lower ionization energies than Mg and, therefore, should be more reactive.

10.23 (a) $2Na(s) + 2H_2O \rightarrow H_2(g) + 2Na^+(aq) + 2OH^-(aq)$
(b) $2Rb(s) + 2H_2O \rightarrow H_2(g) + 2Rb^+(aq) + 2OH^-(aq)$
(c) $Sr(s) + 2H_2O \rightarrow H_2(g) + Sr^{2+}(aq) + 2OH^-(aq)$

10.25 (a) C < N < O < F (b) I < Br < Cl < F

10.27 $2Al(s) + 3Br_2(\ell) \rightarrow 2AlBr_3(s)$ $Zn(s) + Br_2(\ell) \rightarrow ZnBr_2(s)$

10.29 Rust is the product of a direct reaction of iron with oxygen in the presence of moisture to form an iron oxide whose crystals contain water molecules in variable amounts.
$2Fe(s) + 3/2O_2(g) + xH_2O(\ell) \rightarrow Fe_2O_3{\bullet}xH_2O(s)$

10.31 $2Mg(s) + O_2(g) \rightarrow 2MgO(s)$

10.33 (a) $C(s) + O_2(g) \rightarrow CO_2(g)$ (b) $S(s) + O_2(g) \rightarrow SO_2(g)$
(c) $P_4(s) + 5O_2(g) \rightarrow P_4O_{10}(s)$

10.35 (a) $C_9H_{20} + 14O_2 \rightarrow 9CO_2 + 10H_2O$
(b) $2C_2H_4(OH)_2 + 5O_2 \rightarrow 4CO_2 + 6H_2O$
(c) $2(CH_3)_2S + 9O_2 \rightarrow 4CO_2 + 6H_2O + 2SO_2$

10.37 $2C_{20}H_{42} + 21O_2 \rightarrow 40C + 42H_2O$
$2C_{20}H_{42} + 41O_2 \rightarrow 40CO + 42H_2O$
$2C_{20}H_{42} + 61O_2 \rightarrow 40CO_2 + 42H_2O$

10.39 Brønsted-Lowry definitions of acids and bases
Acid: substance that donates a proton (a hydrogen ion H^+) to some other substance.
Base: substance that accepts a proton from an acid.

10.41 (a) H_2SO_4 (b) HSO_4^- (c) H_3O^+
(d) HCl   (e) $HCHO_2$ (or HCOOH)

10.43 Acid-base conjugate pairs (acid written first in each pair)

(a) $HC_2H_3O_2$, $C_2H_3O_2^-$ and H_2O, OH^-
(b) HF, F^- and NH_4^+, NH_3
(c) HSO_4^-, SO_4^{2-} and $H_2PO_4^-$, HPO_4^{2-}
(d) $Al(H_2O)_6^{3+}$, $Al(H_2O)_5OH^{2+}$ and H_2O, OH^-
(e) $N_2H_5^+$, N_2H_4 and H_2O, OH^-
(f) NH_3OH^+, NH_2OH and HCl, Cl^-
(g) OH^-, O^{2-} and H_2O, OH^-
(h) H_2, H^- and H_2O, OH^-
(i) NH_3, NH_2^- and N_2H_4, $N_2H_3^-$
(j) HNO_3, NO_3^- and $H_3SO_4^+$, H_2SO_4

10.45 (a) $2H_2O \rightleftharpoons H_3O^+ + OH^-$

(b) $2NH_3 \rightleftharpoons NH_4^+ + NH_2^-$

(c) $2HCN \rightleftharpoons H_2CN^+ + CN^-$

10.47 $HCO_3^- + H_2O \rightleftharpoons CO_3^{2-} + H_3O^+$ (as an acid)

$HCO_3^- + H_2O \rightleftharpoons H_2CO_3 + OH^-$ (as a base)

10.49 (a) $HClO_3$ (b) HNO_3 (c) H_3PO_4 (d) H_2SO_4
(e) $HClO_3$ (f) $HBrO_3$ (g) H_2Se (h) HBr
(i) PH_3

10.51 CH_3SH This compound and CH_3OH act like weak binary acids. The CH_3SH has a much weaker S-H bond than is the O-H bond in CH_3OH.

10.53 Cl^- is a larger ion than F^- so the HCl bond is much weaker than the HF bond, thus making it more acidic.

10.55 By drawing Lewis structures and then indicating electron shift due to varying electronegativity, one can show an increased negative charge of the central atoms as one goes from $H_2PO_3^-$ to HSO_4^- to ClO_4^-. The shift of electrons away from the oxygen is responsible for $HClO_4$ being a stronger acid than H_2SO_4 which in turn is stronger than H_3PO_4. The question asks for Lewis structures. You should draw them. You may also wish to look at resonance structures of the anions.

10.57 See the answer to Question 10.56. The anion of the stronger acid will be the weaker base. Therefore, HSO_3^- is a stronger base than HSO_4^-.

10.59 Ammonia acts as a Lewis base by donating a pair of electrons to form a covalent bond with H^+ which acts as a Lewis acid.

10.61

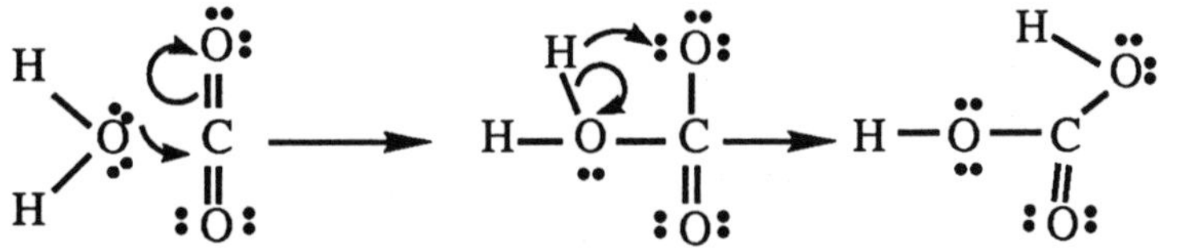

Water acts as a Lewis base by donating a pair of electrons to form a covalent bond with CO_2.

10.63

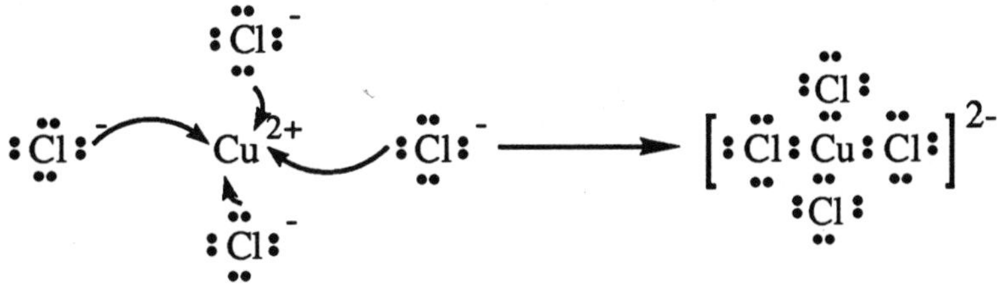

The chloride ions are the Lewis base in the above reaction. Copper ion is the Lewis acid.

10.65 Complexes are also called coordination compounds because they involve coordinate covalent bonds.

10.67 (a) The CN^- displaces the H_2O of $Zn(H_2O)_4^{2+}$ to form $Zn(CN)_4^{2-}$. The CN^- is a stronger ligand (Lewis base) than is water. The Lewis acid is Zn^{2+}.
(b) The Cl^- displaces an NH_3 to form $Pt(NH_3)_3Cl^+$ and free NH_3. The Lewis acid is the Pt^{2+}.

10.69 See the drawing in Section 10.7. EDTA has 6 possible donor atoms. Identify them. EDTA is used in a limited way as a food preservative, in shampoos as a water softener, to prevent clotting of blood samples, and as an antidote for poisoning by heavy metals. It is relatively non-toxic.

10.71 (a) $Cu(NH_3)_4^{2+}$ is a deep blue color
(b) $Cu(H_2O)_4^{2+}$ is a pale blue color
(c) $Co(H_2O)_6^{2+}$ is a reddish-pink color
(d) $Ni(H_2O)_6^{2+}$ is a green color

10.73 (a) AgI_2^- (b) $Ag(NH_3)_2^+$ (c) $Co(C_2O_4)_3^{3-}$
(d) $Co(NO_2)_6^{3-}$ (e) $Cr(EDTA)^-$ (f) $Ni(CN)_4^{2-}$
(g) $Fe(SCN)_6^{3-}$

11 PROPERTIES OF GASES

11.1 Pressure is force per unit area. Pressure is an intensive property. As long as there is a space above the mercury column, the pressures acting along the reference level will be the same regardless of the size and length of the tube. If the diameter of the tube is doubled, there will be four times the weight of Hg acting over four times the area. The ratio of F/A remains unchanged.

11.3 (a) $1.50 \text{ atm} \times \frac{760 \text{ torr}}{\text{atm}} = \mathbf{1{,}140 \text{ torr}}$ or $\mathbf{1.14 \times 10^3 \text{ torr}}$

(b) $785 \text{ torr} \times \frac{1 \text{ atm}}{760 \text{ torr}} = \mathbf{1.03 \text{ atm}}$

(c) $3.45 \text{ atm} \times \frac{101{,}325 \text{ Pa}}{\text{atm}} = 350{,}000 \text{ Pa} = \mathbf{3.50 \times 10^5 \text{ Pa}}$

(d) $3.45 \text{ atm} \times \frac{101.325 \text{ kPa}}{\text{atm}} = \mathbf{350 \text{ kPa}}$

(e) $165 \text{ torr} \times \frac{1 \text{ atm}}{760 \text{ torr}} \times \frac{101{,}325 \text{ Pa}}{\text{atm}} = \mathbf{2.20 \times 10^4 \text{ Pa}}$

(f) $342 \text{ kPa} \times \frac{1 \text{ atm}}{101.3 \text{ kPa}} = \mathbf{3.38 \text{ atm}}$

(g) $11.5 \text{ kPa} \times \frac{1 \text{ atm}}{101.3 \text{ kPa}} \times \frac{760 \text{ torr}}{\text{atm}} = \mathbf{86.3 \text{ torr}}$

11.5 The sketch would be similar to that shown in Figure 11.4(c). The difference in the height of mercury in the two arms would be 25 mm.

11.7 Mercury is often used in barometers and manometers because it has a very high density and, therefore, pressure differences result in convenient height differences. Other reasons for its use are that it is a liquid over a large temperature range and it has a very low vapor pressure.

11.9 (65 mm x 1 torr/mm) + 733 torr = **798 torr**

11.11 (755 mm Hg + 17 mm Hg) x 1 torr/mm Hg = **772 torr**

11.13 $$1 \text{ atm} \times \frac{76.0 \text{ cm Hg}}{\text{atm}} \times \frac{13.6 \text{ g Hg}}{\text{mL Hg}} \times \frac{1 \text{ mL } H_2O}{1 \text{ g } H_2O} \times \frac{1 \text{ in.}}{2.54 \text{ cm}} \times \frac{1 \text{ ft}}{12 \text{ in.}} = 33.9 \text{ ft}$$

(maximum height of a column of water in a sealed column at 1 atm of external pressure)

No, the person will not be able to draw water a height of 35 ft.

11.15 Charles's law is: V/T = constant. Boyle's law is: PV = constant. Gay-Lussac's law is: P/T = constant. PV/T = constant is the combined gas law. If T is constant, then PV = constant, i.e.; Boyle's law. If P is constant, then V/T = constant, i.e.; Charles's law. If V is constant, then P/T = constant, i.e.; Gay-Lussac's law.

11.17 Heating the can causes the pressure of the gas inside to increase. This may cause the can to explode.

11.19 Cooling the gas causes it to contract in volume. This means that a given volume will contain more oxygen.

11.21 Boyle's law states that at a constant temperature, the volume occupied by a fixed quantity of gas is inversely proportional to the applied pressure (or PV = constant). All gases do not always exactly obey Boyle's law. A gas that does would be called an ideal gas.

11.23 $P_iV_i = P_fV_f$ (2.75 atm) x (1.45 L) = [(800/760)atm] x (? L)

? L = **3.79 L**

11.25 $\frac{P_i}{T_i} = \frac{P_f}{T_f}$ $\frac{350\text{ torr}}{(273 + 20)\text{ K}} = \frac{?\text{ torr}}{(273 + 40)\text{ K}}$? torr = **374 torr**

11.27 29 lb /in.^{2} gauge pressure is a total pressure of (14.7 + 29) lb /in.^{2}

65°F = ? °C ? °C = (5/9)(65 - 32) = 18°C

130°F = ? °C ? °C = (5/9)(130 - 32) = 54°C

$\frac{P_i}{T_i} = \frac{P_f}{T_f}$ $\frac{14.7 + 29}{18 + 273°C} = \frac{14.7 + ?\ P}{54 + 273°C}$? P = **34 lb/in.^{2}**

11.29 $\frac{V_i}{T_i} = \frac{V_f}{T_f}$ $\frac{2.0\text{ L}}{(273 + 25)\text{ K}} = \frac{?\text{ L}}{(273 - 28.9)\text{ K}}$? L = **1.6 L**

11.31 $\frac{V_i}{T_i} = \frac{V_f}{T_f}$ $\frac{285\text{ mL}}{(273 + 25)\text{ K}} = \frac{350\text{ mL}}{(273 + ?°C)\text{ K}}$? °C = **93°C**

11.33 $P_iV_i = P_fV_f$ (200 kPa) x (350 cm^3) = (? kPa) x (400 cm^3)

(? kPa) = **175 kPa**

11.35 STP stands for standard temperature and pressure, 0°C (273 K) and one standard atmosphere (760 torr). It is a reference set of conditions.

11.37 $\frac{(450 \text{ torr})(300 \text{ mL})}{(273 + 27) \text{ K}} = \frac{(? \text{ torr})(200 \text{ mL})}{(273 + 20) \text{ K}}$? torr = **659 torr**

11.39 $\frac{(1 \text{ atm})(1 \text{ L})}{(273 + 0) \text{ K}} = \frac{(650/760) \text{ atm}(? \text{ L})}{(273 + 25) \text{ K}}$? L = **1.28 L**

Density = 1.96 g/1.28 L = **1.53 g/L**

11.41 atm liter^2 K^2/mol

11.43 $\frac{0.08206 \text{ L atm}}{\text{mol K}} \times \frac{10^3 \text{ cm}^3}{\text{L}} \times \frac{\text{m}^3}{(100 \text{ cm})^3} \times \frac{1.013 \times 10^5 \text{ Pa}}{\text{atm}}$

$= \mathbf{8.31 \text{ Pa m}^3 \text{ mol}^{-1} \text{K}^{-1}}$

11.45 $245 \text{ mL (STP) } SO_2 \times \frac{1 \text{ mol}}{22.4 \text{ L (STP)}} \times \frac{1 \text{ L}}{1{,}000 \text{ mL}} \times \frac{64.1 \text{ g } SO_2}{\text{mol}} = \mathbf{0.701 \text{ g } SO_2}$

11.47 $\frac{1.96 \text{ g}}{\text{L (STP)}} \times \frac{22.4 \text{ L (STP)}}{\text{mol}} = \mathbf{43.9 \text{ g/mol}}$

11.49 $P = \frac{nRT}{V} = 25.0 \text{ kg} \times \frac{1000 \text{ g}}{\text{kg}} \times \frac{1 \text{ mol } H_2O}{18.0 \text{ g}} \times 0.0821 \text{ L atm mol}^{-1} \text{K}^{-1}$

$\times (273 + 200) \text{ K} \times \frac{1}{1000 \text{ L}} = \mathbf{53.9 \text{ atm}}$ or $\mathbf{41{,}000 \text{ torr}}$

11.51 $$V = \frac{nRT}{P} = \frac{0.234\text{ g}}{17.03\text{ g/mol}} \times 0.08206\text{ L atm mol}^{-1}\text{K}^{-1}$$

$$\times (273 + 30)\text{ K} \times \frac{1}{0.847\text{ atm}} = 0.403\text{ L} = \mathbf{403\text{ mL}}$$

11.53 (a) For C: $$0.482\text{ g } CO_2 \times \frac{1\text{ mol } CO_2}{44.01\text{ g}} \times \frac{1\text{ mol C}}{\text{mol } CO_2} \times \frac{12.01\text{ g C}}{\text{mol}} = 0.132\text{ g C}$$

$$\%\text{ C} = \frac{0.132\text{ g C}}{0.200\text{ g sample}} \times 100\% = \mathbf{66.0\%\text{ C}}$$

For H: $$0.271\text{ g } H_2O \times \frac{1\text{ mol } H_2O}{18.02} \times \frac{2\text{ mol H}}{\text{mol } H_2O} \times \frac{1.01\text{ g H}}{\text{mol}} = 0.0304\text{ g H}$$

$$\%\text{ H} = \frac{0.0304\text{ g H}}{0.200\text{ g sample}} \times 100\% = \mathbf{15.2\%\text{ H}}$$

For N: $$n = \frac{PV}{RT} = \frac{(755/760)(0.0423)}{(0.0821)(273 + 26.5)} = 0.00171\text{ mol } N_2$$

$$0.00171\text{ mol } N_2 \times \frac{2\text{ mol N}}{\text{mol } N_2} \times \frac{14.01\text{ g N}}{\text{mol N}} = 0.0479\text{ g N}$$

$$\%\text{ N} = \frac{0.0479\text{ g N}}{0.2500\text{ g sample}} \times 100\% = \mathbf{19.2\%\text{ N}}$$

(or 100% - 66% - 15.2 % = 18.8%)

(b) $$\text{mol C in 100 g of sample} = \frac{65.8\text{ g C}}{12.01\text{ g/mol}} = 5.48\text{ mol C}$$

$$\text{mol H in 100 g of sample} = \frac{15.2\text{ g H}}{1.01\text{ g/mol}} = 15.0\text{ mol H}$$

$$\text{mol N in 100 g of sample} = \frac{19.2\text{ g N}}{14.01\text{ g/mol}} = 1.37\text{ mol N}$$

Empirical Formula: $C_{5.48/1.37}H_{15.0/1.37}N_{1.37/1.37} = \mathbf{C_4H_{11}N}$

11.55 $1\text{ L} \times 1\text{ atm} \times \frac{101{,}325\text{ Pa}}{\text{atm}} \times \frac{1\text{ N m}^{-2}}{\text{Pa}} \times \frac{1\text{ J}}{1\text{ N m}} \times \frac{0.100000\text{ m}^3}{\text{L}} = \mathbf{101.325\ J}$

$R = 0.0821\text{ L atm mol}^{-1}\text{ K}^{-1} \times 101.325\text{ J L}^{-1}\text{ atm}^{-1} = \mathbf{8.32\ J\ mol^{-1}K^{-1}}$

$R = 8.32\text{ J mol}^{-1}\text{K}^{-1} \times \frac{1\text{ cal}}{4.184\text{ J}} = \mathbf{1.99\ cal\ mol^{-1}\ K^{-1}}$

11.57 (a) $0.00140\text{ mol NO} \times \frac{1\text{ mol }N_2}{2\text{ mol NO}} \times \frac{22.4\text{ L}}{\text{mol}} \times \frac{1{,}000\text{ mL}}{\text{L}} \Leftrightarrow \mathbf{15.7\ mL\ N_2}\text{ (STP)}$

(b) $1.3 \times 10^{-3}\text{ g }H_2 \times \frac{1\text{ mol }H_2}{2.02\text{ g }H_2} \times \frac{1\text{ mol }N_2}{2\text{ mol }H_2} \times \frac{22.4\text{ mL}}{\text{mol}} \times \frac{1{,}000\text{ mL}}{\text{L}}$

$\Leftrightarrow \mathbf{7.2\ mL\ N_2}\text{ (STP)}$

11.59 $10.0\text{ g }HNO_3 \times \frac{1\text{ mol }HNO_3}{63.02\text{ g}} \times \frac{3\text{ mol }NO_2}{2\text{ mol }HNO_3} \times \frac{22.4\text{ L (STP)}}{\text{mol}} \times \frac{760\text{ torr}}{770\text{ torr}}$

$\times \frac{298\text{ K}}{273\text{ K}} \times \frac{1{,}000\text{ mL}}{\text{L}} \Leftrightarrow \mathbf{5{,}740\ mL\ NO_2}$

11.61 $2CO + O_2 \rightarrow 2CO_2$

Moles of CO before reaction =

$\frac{PV}{RT} = \frac{760\text{ atm}}{760} \times \frac{0.500\text{ L}}{1} \times \frac{1\text{ mol K}}{0.0821\text{ L atm}} \times \frac{1}{288\text{ K}} = 0.02115\text{ mol CO}$

Moles of O_2 before reaction =

$\frac{PV}{RT} = \frac{770\text{ atm}}{760} \times .500\text{L} \times \frac{1\text{ mol K}}{0.0821\text{ L atm}} \times \frac{1}{273\text{ K}} = 0.02260\text{ mol }O_2$

(continued)

11.61 (continued)
The limiting reactant is CO.

$$\text{Moles } CO_2 = 0.02115 \text{ mol CO x } \frac{2 \text{ mol } CO_2}{2 \text{ mol CO}} \Leftrightarrow 0.02115 \text{ mol } CO_2$$

$$\text{Volume } CO_2 = \frac{nRT}{P} = 0.02115 \text{ mol x } \frac{0.0821 \text{ L atm}}{\text{mol K}} \text{ x } 301 \text{ K}$$

$$\text{x } \frac{760 \text{ atm}^{-1}}{750} = 0.530 \text{ L or } \mathbf{530 \text{ mL } CO_2}$$

11.63 The sequence of calculations in this problem are: calculate the moles of NO, convert to moles of O_2, and then calculate the volume of the O_2.

$$\frac{PV}{RT} = \frac{750 \text{ atm}}{760} \text{ x } 100 \text{ L x } \frac{\text{mol K}}{0.0821 \text{ L atm}} \text{ x } \frac{1}{773 \text{ K}} = 1.55 \text{ moles NO}$$

$$\text{moles } O_2 = 1.55 \text{ mol NO x } \frac{5 \text{ mol } O_2}{4 \text{ mol NO}} \Leftrightarrow 1.94 \text{ mol } O_2$$

$$\text{vol. } O_2 = \frac{nRT}{P} = 1.94 \text{ mol x } \frac{0.0821 \text{ L atm}}{\text{mol K}} \text{ x } 298 \text{ K x } \frac{1}{0.895 \text{ atm}} = \mathbf{53.0 \text{ L of } O_2}$$

11.65 Dalton's law states that the total pressure exerted by a mixture of gases is equal to the sum of the partial pressures of each gas in the mixture.

11.67

$$P_{N_2} = 300 \text{ torr x } \frac{2.00 \text{ L}}{1.00 \text{ L}} = 600 \text{ torr}$$

$$P_{H_2} = 80 \text{ torr x } \frac{2.00 \text{ L}}{1.00 \text{ L}} = 160 \text{ torr}$$

$$P_T = P_{N_2} + P_{H_2} = 600 + 160 = 760 \text{ torr or } \mathbf{1.00 \text{ atm}}$$

11.69 Oxygen Data

	i	f
P	400 torr	? torr
V	50.0 mL	100 mL
T	60°C	50°C

$$\frac{(400\text{ torr})(50.0\text{ mL})}{(273+60)\text{ K}} = \frac{(?\text{ torr})(100\text{ mL})}{(273+50)\text{ K}}$$

$P_{Oxygen} = ?$ torr = 194 torr

$P_{Nitrogen} = P_{Total} - P_{Oxygen} = 800$ torr - 194 torr = 606 torr

Nitrogen Data

	i	f
P	400 torr	606 torr
V	X mL	100 mL
T	40°C	50°C

$$\frac{(400\text{ torr})(X\text{ mL})}{(273+40)\text{ K}} = \frac{(606)(100\text{ mL})}{(273+50)\text{ K}}$$

X mL of Nitrogen = **147 mL**

11.71 (a) First find the partial pressure of nitrogen $PV = nRT$

$$\text{Pressure of } N_2 = \frac{(0.0020\text{ mol})(0.0821\text{ L atm mol}^{-1}\text{ K}^{-1})(308\text{ K})}{0.200\text{ L}}$$

= 0.253 atm or 192 torr

Use $P_A = X_A P_T$ or $X_A = P_A/P_T$ $X_{Nitrogen} = 192/720 = \mathbf{0.27}$

(b) **190 torr** (only 2 significant figures)

(c) 720 - 190 = 530 torr (only 2 significant figures)

(d) $$n = \frac{PV}{RT} = \frac{(530/760\text{ atm}^{-1})(0.200\text{ L})}{(0.0821\text{ L atm mol}^{-1}\text{K}^{-1})(308\text{ K})} = \mathbf{0.0055\ mol\ of\ O_2}$$

11.73 $P_T = P_{N_2} + P_{CO_2}$ $900 = 800 + ?$ pressure CO_2

Partial pressure CO_2 = 100 torr at 20°C and 500 mL

CO_2 Data

	i	f
P	700 torr	100 torr
V	? V	500 mL
T	303 K	293 K

$$\frac{(700\text{ torr})(?\text{ V})}{(303\text{ K})} = \frac{(100\text{ torr})(500\text{ mL})}{(293\text{ K})}$$

? V = **73.9 mL**

11.75 (a) $n_T = \frac{P_T V}{RT} = \frac{(800/760)(10.0)}{(0.0821)(303)} = \mathbf{0.423\ mol}$

(b) First calculate the moles of N_2 present.

$$0.423 \text{ moles total} = \frac{8.00 \text{ g } CO_2}{44.0 \text{ g } CO_2/\text{mol}} + \frac{6.00 \text{ g } O_2}{32.0 \text{ g } O_2/\text{mol}} + \text{moles } N_2$$

$$\text{moles } N_2 = 0.423 - 0.182 - 0.188 = 0.053$$

$$\text{mol fraction } N_2 = \frac{0.053 \text{ mol } N_2}{0.423 \text{ mol total}} = \mathbf{0.13}$$

$$\text{mol fraction } CO_2 = \frac{0.182 \text{ mol } CO_2}{0.423 \text{ mol total}} = \mathbf{0.430}$$

$$\text{mol fraction } O_2 = \frac{0.188 \text{ mol } O_2}{0.423 \text{ mol total}} = \mathbf{0.444}$$

(c) partial pressure = mol fraction x total pressure
partial pressure N_2 = 0.13 x 800 torr = **100 torr**(only 2 significant figures)
partial pressure CO_2 = 0.430 x 800 torr = **344 torr**
partial pressure O_2 = 0.444 x 800 torr = **355 torr**

(d) $0.053 \text{ mol} \times \frac{28 \text{ g}}{\text{mol}} = \mathbf{1.5\ g\ N_2}$

11.77 Vap. press. of H_2O at 31°C = 33.7 torr

$$n = \frac{PV}{RT} = \frac{(33.7/760)(1)}{(0.0821)(273 + 31)} = 0.00178 \text{ mol } H_2O$$

$$0.00178 \text{ mol } H_2O \times \frac{18.02 \text{ g } H_2O}{\text{mol } H_2O} = \mathbf{0.0320\ g\ H_2O}$$

11.79 **Effusion** is the escape of a gas, under pressure, through a very small opening, while **diffusion** is the spontaneous mixing of two gases placed in the same container.

11.81 (a) $\frac{V_{He}}{V_{Ne}} = \sqrt{\frac{M_{Ne}}{M_{He}}}$

(b) **He**

(c) $\frac{\text{Rate of effusion of He}}{\text{Rate of effusion of Ne}} = \sqrt{\frac{20.2}{4}} =$ **2.25 times faster**

11.83 $\frac{\text{rate of effusion of unknown}}{\text{rate of effusion of } NH_3} = 2.92 = \sqrt{\frac{\text{M.M. } (NH_3)}{\text{M. M. (unknown)}}}$

$$(2.92)^2 = \frac{17.0}{\text{M.M. (unknown)}}$$

$$8.526 = \frac{17.0}{\text{M.M. (unknown)}}$$

M.M. of the unknown = **1.99 g mol^{-1}**

11.85 Pressure arises from the impacts of the molecules of the gas with the walls of the container.

11.87 Raising the temperature increases the average kinetic energy and the average velocity of the molecules of a given gas sample. If the molecules have a greater velocity, they will diffuse more rapidly.

11.89 Raising the temperature increases the pressure because the molecules hit the walls with more force and more often at the higher temperature.

11.91 See Figure 11.13. The curves are not symmetrical because molecules have a lower limit of speed (zero) but virtually no upper limit.

11.93 As gases expand, the average distance of separation of the molecules increases. Since real molecules in a gas attract each other somewhat, moving the molecules further apart requires an increase in potential energy at the expense of kinetic energy. Thus, the average kinetic energy of the molecules decreases, which leads to a decrease in the temperature of the gas.

11.95 Van der Waals subtracted a correction from the value of the measured volume to exclude the volume occupied by the molecules. He added a correction to the measured pressure to correct for the pressure drop in real molecule systems caused by attractions between molecules.

11.97 $$\left(P + \frac{n^2 a}{V^2}\right)(V - nb) = nRT$$

$$\left(P + \frac{1.000^2 \times 5.489}{22.400^2}\right)(22.400 - 1.000 \times 0.06380) = 1.000 \times 0.082057 \times 273.15$$

$$(P + 0.0109)(22.336) = 22.414$$

$$P + 0.0109 = 22.414/22.336 = 1.003$$

$P = 1.003 - 0.0109 =$ **0.992 atm** (ideal gas would have $P = 1$ atm)

11.99 Use the equation $M + HCl \rightarrow 1/2H_2 + MCl$. The equivalent weight of the metal is that mass that will react with 1 mol of H^+ (actually it's that mass that will furnish 1 mol of electrons, but the 2 values are the same). First calculate moles of HCl; then calculate the grams of metal that react with 1 mol of HCl.

$$\text{Mol } H_2 \Leftrightarrow PV/RT = (680 \text{ torr}/760 \text{ torr atm}^{-1})(0.348 \text{ L})/(0.0821 \text{ L atm mol}^{-1} \text{ K}^{-1})(273 + 24) \text{ K} = 1.28 \times 10^{-2} \text{ mol } H_2$$

$$\text{Mol } H^+ = 1.28 \times 10^{-2} \times 2 = 2.56 \times 10^{-2}$$

$$\text{Eq. Wt.} = (0.230 \text{ g}/2.56 \times 10^{-2} \text{ mol}) \times (1 \text{ mol/eq}) = \mathbf{8.98 \text{ g/eq}}$$

The metal is aluminum. $Al + 3HCl \rightarrow 3/2\, H_2 + AlCl_3$

11.101 $$\text{mol } O_2 = (P_{O_2})\frac{V}{RT} = ((740 - 32)/760)(0.250)/(0.0821)(273 + 30) = 9.36 \times 10^{-3} \text{ mol } O_2$$

$$\text{mol } N_2 = (P_{N_2})\frac{V}{RT} = ((780 - 24)/760)(0.300)/(0.0821)(273 + 25) = 12.2 \times 10^{-3} \text{ mol } N_2$$

$$\text{mol}_{(O_2+N_2)} = 9.36 \times 10^{-3} + 12.2 \times 10^{-3} = 21.6 \times 10^{-3} \text{ mol}$$

$$P_{(N_2+O_2)} = (n_{(O_2+N_2)})\frac{RT}{V} = (21.6 \times 10^{-3})(0.0821)(273 + 35)/(0.500 \text{ L}) = 1.09 \text{ atm}$$

$$\text{Total } P_{(N_2+O_2)} = (1.09 \text{ atm} \times 760 \text{ torr/atm}) + 42.2 \text{ torr} = \mathbf{871 \text{ torr}}$$

12 STATES OF MATTER AND INTERMOLECULAR FORCES

12.1 Intermolecular attractions are strong in liquids and solids but not in gases because in liquids and solids the particles are very close while in gases the particles are very far apart.

12.3 Attractions between the positive end of one dipole and the negative end of another are called dipole-dipole attractions. They are weaker in a gas because the molecules are further apart.

12.5 Instantaneous dipole-induced dipole attractions create what we know as London forces.

12.7 Hydrogen bonding forces water molecules into a tetrahedral arrangement about each other which leads to a more "open", less dense structure for ice than for the liquid.

12.9 London forces should increase from helium to neon to argon as the atoms become larger.

12.11 See Figure 12.21. The boiling point comparisons to other compounds are evidence of hydrogen bonding in H_2O, HF and NH_3.

12.13 Density, rate of diffusion, and compressibility are determined primarily by how tightly the molecules are packed.

12.15 At room temperature molecules within solids are very tightly packed and held quite rigidly in place. Therefore, diffusion in solids is virtually nonexistent. At high temperatures the molecules are not as tightly packed and some diffusion can take place.

12.17 Surface tension tends to keep the surface area of the liquid from expanding. When a glass is filled to slightly above its rim, surface tension will keep the liquid from overflowing.

12.19 Draw a diagram that contains the two curves shown in Figure 12.13. One curve is at T_1 and the other is at T_2. It is clear from this diagram that more molecules possess the minimum K.E. for the evaporation at the higher temperature. Therefore, the liquid at that higher temperature will evaporate faster.

12.21 On a dry day water evaporates rapidly because there is little water vapor in the air. The rate of return of H_2O to the clothes is slow compared to the rate of evaporation. When the wind is blowing, the air immediately surrounding the clothes does not have a chance to saturate. Thus, evaporation can continue at a rapid rate. Also the wind can replace some of the heat needed for evaporation.

12.23 A greater fraction of molecules in the warm water have enough energy to escape the surface.

12.25 Increasing the surface area increases the overall rate of evaporation.

12.27 Surface area, temperature, and strengths of the intermolecular attractions determine the rate of evaporation of a liquid.

12.29 Substance X should have the higher boiling point. Substance Y would be less likely to hydrogen bond.

12.31 ΔH_{vap} should increase $PH_3 < AsH_3 < SbH_3$ due to the increased size.

12.33 There are more hydrogen bonds in water than in HF because water has 2 hydrogen atoms whereas HF has only one.

12.35 The source of energy in a thunderstorm is the heat of vaporization that is liberated when H_2O condenses.

12.37 $\frac{\Delta Hvap}{B.P.} = \text{const.}$ For CH_4: $\frac{9.20 \text{ kJ/mol}}{(273 - 161)\text{K}} \times 10^3 \text{ J/kJ} = \mathbf{82.1 \text{ J mol}^{-1} \text{K}^{-1}}$

For C_2H_6: **76 J mol^{-1} K^{-1}**

For C_3H_8: **74.5 J mol^{-1} K^{-1}**

For C_4H_{10}: **81.7 J mol^{-1} K^{-1}**

For C_6H_{1}[illegible]: **83.0 J mol^{-1} K^{-1}**

For C_8H_{18}: **85.2 J mol^{-1} K^{-1}**

For $C_{10}H_{22}$: **82.7 J mol^{-1} K^{-1}**

(ΔH_{vap} is directly proportional to B.P. because they both depend on intermolecular attractions.)

12.39 $35.0 \text{ g} \times \frac{1 \text{ mol}}{78.12 \text{ g}} \times \frac{9.92 \text{ kJ}}{\text{mol}} \Leftrightarrow \mathbf{4.44 \text{ kJ}}$

12.41 K.E. = $1/2mv^2$ = 1/2(68.2 kg)(10.0 mi/hr)^2 (1 hr/60 min)^2 (1 min/60 s)^2

(5280 ft/mi)^2 (12 in./ft)^2 (2.54 cm/in.)^2 (1 m/100 cm)^2 (1 J s^2/kg m^2)

= 6.81×10^2 J $\Delta H_{fus.}(H_2O)$ = 5.98 kJ/mol (Table 12.3)

$$6.81 \times 10^2 \text{J} \times \frac{1 \text{ mol}}{5.98 \text{ kJ}} \times \frac{1 \text{ kJ}}{1{,}000 \text{ J}} \times \frac{18.02 \text{ g}}{\text{mol}} = \mathbf{2.05 \text{ g}}$$

12.43 Energy absorbed by the benzene = $\Delta H_{fus.}$ x $mol_{benzene}$

= 9.92 kJ/mol x 10.0 g /(78.06 g/mol) = 1.27 kJ

1.27 kJ = energy also released by the H_2O = sp. heat x g x change in temp.

1.27 kJ = (4.184 J/g °C) x (50.0 g) x (30.0 - ?)°C x (1.00 kJ/1000 J)

6.07 = (30.0 - ?) ? = **23.9°C**

12.45 The "equilibrium vapor pressure" is the pressure exerted by the gaseous molecules of a substance above the liquid in a closed system at equilibrium.

12.47 Ethyl alcohol

12.49 The equilibrium vapor pressure depends on the rate of evaporation of the liquid, which is determined by the fraction of molecules having enough K.E. to escape. If the attractive forces are large, this fraction is small.

12.51 Vapor pressure is the equilibrium pressure of the gas in the equilibrium, liquid $\rightleftharpoons$ gas. The value of the equilibrium gas pressure is independent of volume of liquid or gas. Increased surface area of the liquid will increase the rate at which equilibrium can be attained but not the value at equilibrium. A larger surface area will allow more molecules to enter the gaseous phase but at the same time, if equilibrium exists, more will be condensing to the liquid phase.

12.53 A humidity of 100% implies air saturated with H_2O vapor which exerts a partial pressure equal to the vapor pressure of water at that temperature. Vapor pressure increases with temperature; therefore, the amount of water in a given volume of saturated air also increases.

12.55 The critical temperature is the highest temperature at which the substance can exist as a liquid regardless of the pressure. The critical pressure is the vapor pressure of a substance at its critical temperature.

12.57 Below its critical temperature, -267.8°C (Table 12.2)

12.59 The water is able to evaporate more quickly if the saturated atmosphere above the coffee is slowly blown away. This increased rate of evaporation allows the coffee to cool more quickly.

12.61 The vapor pressure of a solid increases with increased temperature.

12.63 Le Chatelier's principle states that "when a system in a state of dynamic equilibrium is disturbed by some outside influence that upsets the equilibrium, the system responds by undergoing a change in a direction that reduces the disturbance and, if possible, brings the system back to equilibrium."

12.65 It will shift the equilibrium to the left since the solid will occupy much less space.

12.67 **~88°C**

12.69 The stronger the intermolecular attractive forces, the more difficult it is for molecules to break away from the liquid and enter the gaseous state. To make a compound with strong intermolecular forces boil, one has to provide more kinetic energy in the form of a higher temperature.

12.71 The temperature at which solid $\rightleftharpoons$ liquid exists is called either the freezing point or the melting point.

12.73 Fusion is the process of melting.

12.75 Crystalline solids have highly regular, symmetrical shapes. They possess faces that intersect each other at characteristic interfacial angles.

12.77 A lattice is a regular or repetitive pattern of points or particles. A unit cell is the grouping of particles that is repeated throughout the solid and generates the entire lattice. We can create an infinite number of chemical structures by simply varying the chemical environment about each point in a lattice.

12.79 The quantities a, b & c, corresponding to the edge lengths of the cell and α, β & γ, corresponding to the angles at which the edges intersect one another, describe a particular lattice.

12.81 (a) Its lattice arrangement is a face-centered cubic lattice of chloride ions with sodium ions at the center of each edge and in the center of the unit cell.

(b) Cl^- 8 corners x 1/8Cl^- per corner = 1 Cl^-

6 faces x 1/2Cl^- per face = 3 Cl^-

Total = 4 Cl^-

Na^+ 12 edges x 1/4Na^+ per edge = 3 Na^+

1 center x 1 Na^+ per center = 1 Na^+

4 Na^+

Na_4Cl_4 or **4 Formula Units per Unit Cell**

(c) **No.** The cations and anions would need to have the same magnitude of charge if all the corners and faces are to be occupied (i. e., anion/cation ratio of one-to-one).

12.83 $2d \sin \theta = n\lambda$ $n = 1$ $\lambda = 141$ pm

(a) $2d \sin 20.0° = 1 \times 141$

$2 \times d \times 0.342 = 141$ $d = 141/0.684 =$ **206 pm**

(b) $2d \sin 27.4° = 1 \times 141$ $d = 141/0.920 =$ **153 pm**

(c) $2d \sin 35.8° = 1 \times 141$ $d = 141/1.170 =$ **121 pm**

12.85 (Draw a picture of the unit cell). In the body-centered cubic unit cell there is a sphere at each corner and in the very center. The corner spheres each touch the center sphere but they do not touch each other. The distance from one corner to the corner diagonally through the unit cell is 4 radii or 4 r. From the unit cell two right triangles can be constructed. The first involves 2 edges (E) and the diagonal across the face (DF). The other involves the unit cell diagonal (DC), an edge (E), and the diagonal across the face (DF). In the first triangle we can use the known edge length (E) to calculate the length of the diagonal across the face of the unit cell (DF). Then in the second triangle we can use the now known length of the diagonal across the face (DF) and the edge length (E) to calculate the unit cell diagonal (DC). Knowing the unit cell diagonal (DC) is 4r, we then have the radius of the atom.

First Triangle $(E)^2 + (E)^2 = (DF)^2$

$(288.4)^2 + (288.4)^2 = (DF)^2$ DF = 407.86 pm

Second Triangle $(E)^2 + (DF)^2 = (DC)^2$

$(288.4)^2 + (407.86)^2 = (DC)^2$ $249{,}500 = (DC)^2$

DC = 499.5 pm 4r = 499.5 pm **r = 124.9 pm**

12.87 A diagonal (DF) drawn across the face of the unit cell forms a right triangle with the edges of the unit cell (E) as the two sides of the the triangle. The diagonal (DF) is 4 radii.

$(4r)^2 = (DF)^2 = (E)^2 + (E)^2$ $(4 \times 143)^2 = 2(E)^2$

$E^2 = 163{,}600$ **E** (or length) = **404 pm**

12.89 Unit cell edge length = 658 pm

658 pm = 2r (Rb^+) + 2r (Cl^-) 658 pm = 2r(Rb^+) + (2 x 181 pm)

r (Rb^+) = 148 pm

148 pm or **1.48 angstroms**

12.91 Volume of sphere = $4/3\pi\, r^3 = 4/3\pi\,(50\text{ pm})^3 = 5.236 \times 10^5\text{ pm}^3$

(a) volume of primitive cubic unit cell (See Problem 12.90).

$V = (2r)^3 = (2 \times 50\text{ pm})^3 = 1.0 \times 10^6\text{ pm}^3$

Vacant space = $1.0 \times 10^6\text{ pm} - 5.24 \times 10^5\text{ pm}^3 = 4.8 \times 10^5\text{ pm}^3$

$$\frac{4.8 \times 10^5}{1.0 \times 10^6} \times 100\% = \mathbf{48\%\ vacant}$$

(b) For a discussion of the body-centered cubic unit cell see Problem 12.87.
$(4r)^2 = (DC)^2 = (E)^2 + (E)^2 + (E)^2$ $(4 \times 50\text{ pm})^2 = 3(E)^2$

$E = 115\text{ pm}$ $V = (115\text{ pm})^3 = 1.52 \times 10^6\text{ pm}^3$ (2 spheres per unit cell)

Vacant space = $1.52 \times 10^6\text{ pm}^3 - (2 \times 5.24 \times 10^5\text{ pm}^3) = 4.7 \times 10^5\text{ pm}^3$

$$\frac{4.7 \times 10^5}{1.52 \times 10^6} \times 100\% = \mathbf{31\%\ vacant}$$

(c) For the face-centered cubic unit cell see Problem 12.86.
$(4r)^2 = (DF)^2 = (E)^2 + (E)^2$ $(4 \times 50)^2 = 2(E)^2$

$E = 141\text{ pm}$

$V = (141\text{ pm})^3 = 2.80 \times 10^6\text{ pm}^3$ (Four spheres per unit cell)

Vacant space = $2.80 \times 10^6 - (4 \times 5.24 \times 10^5) = 7.04 \times 10^5\text{ pm}^3$

$$\frac{7.04 \times 10^5}{2.80 \times 10^6} \times 100\% = \mathbf{25.1\%\ vacant}$$

12.93 Volume of the cube = $(412.3\ pm)^3 = 7.009 \times 10^7\ pm^3$

Atoms per unit cell: if face-centered = 4CsCl (See 12.85)

if body-centered = 2CsCl

Calculated density of the face-centered unit cell =

$$\frac{4 \text{ formula units CsCl}}{7.009 \times 10^7 pm^3} \times \frac{168.4\ g}{mole} \times \frac{1 \text{ mole}}{6.022 \times 10^{23} \text{ form. units}} \times \frac{(10^{10} pm)^3}{cm^3}$$

$= \mathbf{15.96\ g/cm^3}$

Not very close to the known value of 3.99 g/cm³

Calculated density of the body-centered unit cell =

$$\frac{2 \text{ formula units CsCl}}{7.009 \times 10^7 pm^3} \times \frac{168.4\ g}{mole} \times \frac{1 \text{ mole}}{6.022 \times 10^{23} \text{ form. units}} \times \frac{(10^{10} pm)^3}{cm^3}$$

= **7.979 g/cm³ Not very close to the known value of 3.99 g/cm³**
(Calculation of the density of the primitive cubic cell yields 3.99 g/cm^3, the value of CsCl.)

12.95 vol/unit cell = $[546.26\ pm \times (10^{-10}\ cm/pm)]^3 = 1.6300 \times 10^{-22}\ cm^3$

$$\frac{1.6300 \times 10^{-22}\ cm^3}{\text{unit cell}} \times \frac{3.180\ g}{cm^3} \times \frac{1\ mol}{78.08\ g} \times \frac{6.022 \times 10^{23} \text{ form. units}}{mol}$$

= **3.998 formula units/unit cell ≈ 4**

12.97 Please see Table 12.5.

12.99 (a) molecular

(b) ionic

(c) ionic

(d) metallic

(e) covalent

(f) molecular

(g) ionic

12.101 Boron is probably a covalent solid based on the data given.

12.103 OsO_4 is probably a molecular solid.

12.105 Prepare a heating curve like that in Figure 12.38 and give the actual values of T_f, T_b, ΔH_{fus} and ΔH_{vap}.

12.107 Glass is an amorphous solid. The disorientation of the molecules gives rise to a melting point range.

12.109 All the heat that is added is used to increase the potential energy of the molecules being converted to the gas phase. Therefore, the average kinetic energy and the temperature of all the molecules (both gaseous and liquid) remains constant until all the molecules have become gaseous.

12.111

At 22°C

State	Pressure (torr)
vapor	up to 160
vapor-liquid	160
liquid	160 to 250
solid-liquid	250
solid	250 to 1,000

At 10°C

State	Pressure (torr)
vapor	up to 75
solid-vapor	75
solid	75 to 1,000

12.113 The density of the solid is greater than that of the liquid because the solid-liquid line leans to the right (connecting the triple point, 20° - 150 torr, and the melting point, 25° - 760 torr).

12.115 The triple point of I_2 occurs at a temperature above that to which it was heated and at a pressure above atmospheric pressure.

12.117 For (b) boiling point, (c) heat of vaporization, (d) surface tension, (f) heat of sublimation, and (g) critical temperature to be arranged from highest value to lowest value, the compounds would follow the order: propylene glycol, isopropyl alcohol, acetone, methyl ethyl ether, and butane. For (a) vapor pressure and (e) rate of evaporation, the order would be reversed with butane having the highest value.

12.119 AY + YB = extra distance traveled by more penetrating ray = $n\lambda$; XY = d

$$\sin\theta = \frac{YB}{XY} = \frac{YB}{d} \text{ and } \sin\theta = \frac{AY}{XY} = \frac{AY}{d} \text{ or } YB = d\sin\theta \text{ and } AY = d\sin\theta$$

therefore, $AY + YB = d\sin\theta + d\sin\theta = n\lambda$

$$2d\sin\theta = n\lambda$$

13 PHYSICAL PROPERTIES OF COLLOIDS AND SOLUTIONS

13.1 Pure substances, unlike mixtures, have constant composition.

13.3 **From 1nm to 1,000 nm**
(a) **1×10^{-6} mm to 1×10^{-3} mm**
(b) **3.9×10^{-8} in. to 3.9×10^{-5} in.**

13.5 By spinning a sample in a centrifuge, the centrifugal force thus produced behaves as a very powerful artificial gravity and drives the suspended particles to the bottom of the container.

13.7

	Dispersing phase	Dispersed phase	Kind of colloid
(a) Styrofoam	solid	gas	solid foam
(b) cream	liquid	liquid	emulsion
(c) lard	solid	liquid	solid emulsion
(d) jelly	liquid	solid	gel, sol
(e) liq. rubber cement	liquid	solid	sol, gel

13.9 Test for the Tyndall effect.

13.11 First AgCl forms and begins to form dispersed particles. The excess Cl^- adsorbs on the AgCl particles and acts to prevent them from growing too large to remain suspended. As more Ag^+ is added, the electrical charge is neutralized by the formation of more AgCl and the particles grow in size and will settle out of solution.

13.13 Solid, liquid and gaseous solutions are possible.

13.15 Substitutional solid solutions exist when atoms, molecules, or ions of the solute replace particles of the solvent in the crystalline lattice. Brass is an example.

13.17 An interstitial solid solution would be more likely to exist under the conditions described in this question.

13.19 All concentration units are ratios (fractions).

13.21 45.0 g of benzene = 45.0 g/78.06 g/mol = 0.576 mol benzene

80.0 g of toluene = 80.0 g/92.15 g/mol = 0.868 mol toluene

(a) $\text{weight \% toluene} = \dfrac{80.0\text{ g}}{80.0\text{ g} + 45.0\text{ g}} \times 100\% = \mathbf{64.0\%\ toluene}$

$\text{weight \% benzene} = \dfrac{45.0\text{ g}}{80.0\text{ g} + 45.0\text{ g}} \times 100\% = \mathbf{36.0\%\ benzene}$

(b) $X_{\text{toluene}} = \dfrac{0.868\text{ mol toluene}}{0.868\text{ mol} + 0.576\text{ mol}} = \mathbf{0.601}$

(continued)

13.21 (continued)

$$X_{benzene} = \frac{0.576 \text{ mol benzene}}{0.868 \text{ mol} + 0.576 \text{ mol}} = \mathbf{0.399}$$

(c) $$m = \frac{0.576 \text{ mol benzene}}{0.0800 \text{ kg toluene}} = \mathbf{7.20\ m}$$

13.23 40.0 mol of $H_2O \Leftrightarrow$ 40.0 mol x 18.02 g/mol or 721 g

0.30 mol of $CuCl_2 \Leftrightarrow$ 0.30 mol x 134.5 g/mol or 40 g

(a) X of $CuCl_2$ = 0.30/(0.30 + 40.0) = **0.0074**

(b) m of $CuCl_2$ = 0.30 mol/0.721 kg H_2O = **0.42 m**

(c) weight % $CuCl_2$ = [40 g/(40 + 721) g] x 100% = **5.3%**

13.25 $$X_{alcohol} = 0.250 = \frac{0.250 \text{ mol alcohol}}{0.250 \text{ mol alc.} + ?\text{ mol } H_2O}$$

? mol H_2O= 0.750 mol

$$\text{wt. \% alc.} = \frac{0.250 \text{ mol x } 60.1 \text{ g mol}^{-1}}{(0.250 \text{ mol x } 60.1 \text{ g mol}^{-1}) + (0.750 \text{ mol x } 18.02 \text{ g mol}^{-1})} \text{ x } 100\%$$

$$= \frac{15.02 \text{ g}}{28.54 \text{ g}} \text{ x } 100\% = \mathbf{52.6\%}$$

$$m\,(\text{alc}) = \frac{0.250 \text{ mol alc.}}{0.01352 \text{ kg } H_2O} = \mathbf{18.5\ m}$$

13.27 $6.25\ m = \dfrac{6.25\ \text{mol NaCl}}{1.000\ \text{kg } H_2O}$

$$X_{NaCl} = \frac{6.25\ \text{mol NaCl}}{6.25\ \text{mol NaCl} + (1{,}000\ \text{g } H_2O/\ 18.02\ \text{g mol}^{-1})} = \frac{6.25}{6.25 + 55.49} = \mathbf{0.101}$$

$$\omega_{NaCl} = \frac{6.25\ \text{mol NaCl x } 58.44\ \text{g mol}^{-1}}{(6.25\ \text{mol NaCl x } 58.44\ \text{g mol}^{-1}) + 1{,}000\ \text{g } H_2O} = \frac{365\ \text{g}}{365\ \text{g} + 1{,}000\ \text{g}} = \mathbf{0.267}$$

13.29 (a) $M = \dfrac{150 \times 10^{-3}\ \text{g}/24.3\ \text{g mol}^{-1}}{1.00\text{L}} = \mathbf{0.00617\ M\ Mg^{2+}}$

(b) $m = \dfrac{150 \times 10^{-3}\ \text{g}/24.3\ \text{g mol}^{-1}}{1\ \text{kg} - 150 \times 10^{-6}\text{kg}} = \dfrac{6.173 \times 10^{-3}\ \text{mol}}{0.99985\ \text{kg}} = \mathbf{0.00617\ m\ Mg^{2+}}$

Note: For very dilute solutions molarity and molality are equal to 2 or 3 significant figures.

13.31 $2.25\ m\ NH_4NO_3\ \text{soln.} = \dfrac{2.25\ \text{mol } NH_4NO_3}{1.00\ \text{kg } H_2O}$

$$\text{Weight \% } NH_4NO_3 = \frac{2.25\ \text{mol } NH_4NO_3 \text{ x } 80.06\ \text{g mol}^{-1}}{2.25\ \text{mol x } 80.06\ \text{g mol}^{-1} + 1{,}000\ \text{g } H_2O} \text{ x } 100\%$$

$$= \frac{180.1}{180.1 + 1{,}000} = \mathbf{15.3\%\ NH_4NO_3}$$

$$X_{NH_4NO_3} = \frac{2.25\ \text{mol } NH_4NO_3}{2.25\ \text{mol } NH_4NO_3 + 1{,}000\ \text{g } H_2O/18.02\ \text{g mol}^{-1}}$$

$$= \frac{2.25}{2.25 + 55.49} = \mathbf{0.0390}$$

$$X_{H_2O} = \frac{55.49}{2.25 + 55.49} = \mathbf{0.9610}$$

13.33 $m = \dfrac{222.6\ g/62.08\ g\ mol^{-1}}{0.2000\ kg\ H_2O} = \mathbf{17.93\ m}$

$$M = \frac{222.6\ g/62.08\ g\ mol^{-1}}{(200.0\ g + 222.6\ g)(1\ mL/1.072\ g)(1\ L/1{,}000\ mL)} = 9.096\ \frac{mol}{L} = \mathbf{9.096\ M}$$

13.35 1 kg of H_2O will contain the number of moles of the ions listed in the table. That number of moles will weigh: (0.566 mol x 35.45 g mol^{-1}) + (0.486 mol x 22.99 g mol^{-1}) + (0.055 mol x 24.3 g mol^{-1}) + (0.029 mol x 96.1 g mol^{-1}) + (0.011 mol x 40.1 g mol^{-1}) + (0.011 mol x 39.1 g mol^{-1}) + (0.002 mol x 61 g mol^{-1}) = 36.35 g of ions

$$\text{Mass of } Cl^- = \frac{0.566\ mol\ Cl^-}{1\ kg\ H_2O} \times \frac{35.45\ g\ Cl^-}{mol} \times \frac{1\ kg\ H_2O}{(1{,}000\ g\ H_2O + 36.35\ g\ ions)}$$

$$\times \frac{1024\ g\ soln.}{L} \times 3.78\ L\ soln. = \mathbf{74.9\ g\ Cl^-}$$

$$\text{Mass of } Na^+ = \frac{0.486\ mol\ Na^+}{1\ kg\ H_2O} \times \frac{22.99\ g\ Na^+}{mol} \times \frac{1\ kg\ H_2O}{(1{,}000\ g\ H_2O + 36.35\ g\ ions)}$$

$$\times \frac{1024\ g\ soln.}{L} \times 3.78\ L\ soln. = \mathbf{41.7\ g\ Na^+}$$

Mass of Mg^{2+} = 0.055 x 24.3 x 3.73 = **5.0 g Mg^+**
Mass of SO_4^{2-} = 0.029 x 96.1 x 3.73 = **10 g SO_4^{2-}**
Mass of Ca^{2+} = 0.011 x 40.1 x 3.73 = **1.6 g Ca^{2+}**
Mass of K^+ = 0.011 x 39.1 x 3.73 = **1.6 g K^+**
Mass of HCO_3^- = 0.002 x 61 x 3.73 = **0.5 g HCO_3^-**

Total Mass of ions = **135.3 g** or **135 g**

13.37 Since the gas molecules are in constant, random motion, they mix in a short period of time. The random motion of the gas molecules will yield greater disorder if possible and, in this case, that will result in spontaneous mixing.

13.39 In the case of the mixing of hexane and salt, the salt has very strong ionic forces that would need to be broken before it could be distributed throughout the hexane. The tendency toward disorder alone cannot overcome the strong attraction within the solid salt. Therefore, a salt-hexane solution cannot be formed.

13.41 Ammonia is quite soluble because it can form solute-solvent forces of attraction with water (hydrogen bonds) that are comparable to the strength of the solvent-solvent forces of attraction in water. H_2 and O_2 are incapable of forming such strong solute-solvent forces of attraction with water.

13.43 Substances that exhibit similar intermolecular attractive forces tend to be soluble in one another.

13.45 Methyl alcohol molecules surround H_2O molecules resulting in micelle-like particles that are soluble in the gasoline.

13.47 When dissolving an ionic compound the water molecules surround the ions "insulating" their charges from each other.

13.49 They do not form precipitates with the cations found in hard water.

13.51 Positive

13.53 When hexane dissolves in ethanol, some of the hydrogen bonding between ethanol molecules will be disrupted. The energy needed to break hydrogen bonds in ethanol is not regained when ethanol and hexane are mixed since no strong forces exist between ethanol and hexane.

13.55 The heat of solution is equal to the difference between the lattice energy and the hydration energy. If the lattice energy is greater, a net input of energy is required; therefore, the solution process is endothermic.

13.57 $AlCl_3$ probably has a large value for its hydration energy.

13.59 $\Delta H_{soln.}$ ($AlCl_3$) = -321 kJ/mol

$$\frac{-321\text{ kJ}}{\text{mol}} \times 10.0\text{ g} \times \frac{1\text{ mol}}{133.3\text{ g}} = -24.1\text{ kJ}$$ or **24.1 kJ liberated**

13.61 KI would become more soluble with an increase in temperature. Addition of heat favors an endothermic process. Using Le Chatelier's principle, heat is a reactant in this process and solution is a product. Increased heat will yield increased solution.

13.63 Since the solution process for a gas in a liquid is nearly always exothermic, the reverse reaction is endothermic and gases are less soluble as the temperature is raised.

13.65 Pressure only has an appreciable effect on equilibria where sizable volume changes occur. When a liquid or solid dissolves in a liquid, only very small changes in volume occur.

13.67 $p_g = C_g/k_g = 0.0478\text{ g L}^{-1}/6.50 \times 10^{-5}\text{ g L}^{-1}\text{ torr}^{-1} = 735\text{ torr}$

$P_{total} = p_g + p_{water} = 735 + 23.8 =$ **759 torr**

13.69 $(5.34 \times 10^{-5}\text{ g L}^{-1}\text{ torr}^{-1})(0.20 \times (760 - 24)\text{torr}) = 7.86 \times 10^{-3}\text{ g/L}$

$7.86 \times 10^{-3}\text{ g/L} \times 1\text{L} =$ **7.9×10^{-3} g**

13.71 When the vapor pressure of a mixture is greater than that predicted, it is said to exhibit a positive deviation from Raoult's law; conversely, when a solution gives a lower vapor pressure than we would expect from Raoult's law, it is said to show a negative deviation.

13.73 Raoult's Law states that $P_{soln.} = X_{solvent} P^{\circ}_{solvent}$

$P^{\circ}_{solvent} = 93.4$ torr

$$X_{solvent} = \frac{1{,}000 \text{ g benzene}/78.06 \text{ g mol}^{-1}}{(1{,}000 \text{ g benzene}/78.06 \text{ g mol}^{-1}) + (56.4 \text{ g } C_{20}H_{42}/283 \text{ g mol}^{-1}}$$

$$= \frac{12.81}{13.01} = 0.985$$

$P_{soln.} = 0.985 \times 93.4$ torr = **92.0 torr**

13.75 $P_T = X_A P^{\circ}_A + X_B P^{\circ}_B$

$$P_T = \frac{25.0 \text{ g}/100.1 \text{ g mol}^{-1}}{(25.0 \text{ g}/100.1 \text{ g mol}^{-1}) + (35.0 \text{ g}/114.1 \text{ g mol}^{-1})} \times 791 \text{ torr}$$

$$+ \frac{35.0 \text{ g}/114.1 \text{ g mol}^{-1}}{(25.0 \text{ g}/100.1 \text{ g mol}^{-1}) + (35.0 \text{ g}/114.1 \text{ g mol}^{-1})} \times 352 \text{ torr}$$

$$= \frac{0.250}{0.250 + 0.307} \times 791 \text{ torr} + \frac{0.307}{0.250 + 0.307} \times 352 \text{ torr}$$

= (0.449 x 791 torr) + (0.551 x 352 torr) = **549 torr**

13.77 $P_{soln.} = X_A P°_A + X_B P°_B$

$$137 \text{ torr} = \frac{400 \text{ g}/154 \text{ g mol}^{-1}}{(400 \text{ g}/154 \text{ g mol}^{-1}) + (43.3 \text{ g}/? \text{ mol}^{-1})} \text{ x } 143 \text{ torr}$$

$$+ \frac{43.3 \text{ g}/ ? \text{ mol}^{-1}}{(400 \text{ g }/154 \text{ g mol}^{-1}) + (43.3 \text{ g }/? \text{ g mol}^{-1})} \text{ x } 85 \text{ torr}$$

$$137 = \frac{(2.597)(143)}{2.597 + 43.3/?} + \frac{(43.3/?)(85)}{2.597 + 43.3/?}$$

$$2.597 + 43.3/? = \frac{371.4}{137} + \frac{3681/?}{137} \qquad 43.3/? - \frac{3681/?}{137} = 0.1139$$

$$5{,}932/? - 3{,}681/? = 15.604 \qquad ? = \mathbf{144 \text{ g/mol}}$$

13.79 $$X_{water} = \frac{100 \text{ g}/18.02 \text{ g mol}^{-1}}{(100 \text{ g}/18.02 \text{ g mol}^{-1}) + (150 \text{ g}/92.11 \text{ g mol}^{-1})} = 0.773$$

$91.1 \text{ torr} = 0.773\ P°_{water}$ $\qquad P°_{water} = 118 \text{ torr}$

118 torr = **55°C** (From Table 11.2)

13.81 Approximately three times for Figure 13.22 and 4 or 5 times for Figure 13.23.

13.83 Draw a diagram like that in Figure 13.23. If water is on the left and butyl alcohol on the right, the lines will start at 100°C (boiling point of water), decrease to 92.4°C (boiling point of the azeotrope, which is a mole fraction of water of 0.716), and then increase to 117.8°C at the right axis (the boiling point of butyl alcohol). These substances show positive deviation from ideality, i. e., minimum-boiling azeotrope.

13.85 In the presence of a solute, the rate of freezing at a particular temperature is decreased because fewer solvent particles are in contact with the solid. The rate of melting, however, is the same since no solute is incorporated in the solid solvent to re-establish equilibrium. The temperature must be lowered so that the solvent freezes faster from the solution and the solvent melts more slowly from the solid until the rates of these two processes become equal.

13.87 $\Delta T_f = K_f m$

$$0.307°\,C = (5.12°C\,m^{-1})\left(\frac{3.84\ g/MM}{0.500\ kg\ benzene}\right)$$

MM = **128** Empirical formula is given as C_4H_2N

Therefore, the molecular formula based upon calculated MM must be twice the empirical formula: $\mathbf{C_8H_4N_2}$

13.89 $\Delta Tf = Kf\,m$ $$0.750°\,C = (1.86°C\,m^{-1})\left(\frac{?\ g/180.2\ g\ mol^{-1}}{0.150\ kg\ H_2O}\right)$$

? = **10.9** g added

$$\Delta T_b = K_b\ m = (0.51°C\,m^{-1})\left(\frac{10.9\ g/180.2\ g\ mol^{-1}}{0.150\ kg}\right) = 0.21°C$$

B.P. = 100.00 + 0.21 = **100.21°C**

13.91 0.075 = fraction dissociated, (assume $MX \rightarrow M^+ + X^-$)

1.000 - 0.075 = fraction undissociated = 0.925

Total = [(0.925) x 0.100 m] + [(2 x 0.075) x 0.100 m] = 0.108 m

$\Delta T_f = (1.86°C\ m^{-1})\ (0.108\ m) = 0.201°C$

F.P. = 0.000°C - 0.201°C = **-0.201°C**

13.93 In dialysis small ions, small molecules, and solvent are allowed to pass through a membrane, but in osmosis only solvent is allowed through the membrane.

13.95 Solutions that have the same osmotic pressure are called isotonic solutions. During intravenous feeding, the solute concentration must be carefully controlled to prevent excessive movement of fluid into or out of the cells.

13.97 $\pi = MRT$ $$\frac{3.74 \text{ torr}}{760 \text{ torr atm}^{-1}} = \frac{(0.400 \text{ g/MM})}{1.00 \text{ L}}(0.0821 \text{ L atm mol}^{-1}\text{K}^{-1})(300 \text{ K})$$

MM = **2,000** (3 sign. figures)

13.99 $$\Delta T_f = m\,K_f = 0.10 \text{ m } CaCl_2 \times \frac{3 \text{ m ions}}{1 \text{ m salt}} \times 1.86°\text{C m}^{-1} = 0.56°\text{C}$$

F.P. = 0.00°C - 0.56 = **-0.56°C** (if completely dissociated)

13.101 $\pi = MRT = 2 \times 0.010 \text{ M} \times (0.0821 \text{ atm M}^{-1} \text{ K}^{-1})(298 \text{ K})(760 \text{ torr/atm})$

= **370 torr** (2 sign. figures)

13.103 $\Delta T_f = m\,K_f$

$\Delta T_f = (0.10 \text{ m})(1.86°\text{C m}^{-1}) = 0.19°\text{C}$ (calculated as a nonelectrolyte solution)

$$i = \frac{(\Delta T_f) \text{ measured}}{(\Delta T_f) \text{ calculated as nonelectrolyte}}$$

i = 1.21 (from Table 13.6)

$$1.21 = \frac{(\Delta T_f) \text{ measured}}{0.19}$$

$\Delta T_f = 0.23°\text{C}$

F.P. = 0.00 - 0.23 = **-0.23°C**

13.105 $Al_2(SO_4)_3$

13.107 Calculated freezing point: $\Delta T_f = (0.130\ m)(1.86°C\ m^{-1})$

$\Delta T_f = 0.242$ calculated as nonelectrolyte

$$i = \frac{(\Delta T_f)\ \text{measured}}{(\Delta T_f)\ \text{calculated}} = \frac{0.72}{0.242} = 2.98 = 3.0$$

This i value is consistent with: $Hg_2Cl_2(s) + H_2O \rightarrow Hg_2^{2+}(aq) + 2Cl^-(aq)$
(in which the mercury(I) remains a dimer).

13.109 Assuming that ethylene glycol is nonvolatile, which it is very close to being at the temperatures in this problem:

$$\Delta T_b = mK_b = \left(\frac{10.0\ g/62.08\ g\ mol^{-1}}{0.0900\ kg}\right) 0.51 = 0.913°C$$

B.P. = **100.91°C**

$$\Delta T_f = mK_f = \left(\frac{10/62.08}{0.0900}\right) 1.86 = 3.33°C$$

F.P. = **-3.33°C**

$$\text{V.P.} = 760\ \text{torr}\left(\frac{90.0\ g/18.01\ g\ mol^{-1}}{(90.0\ g/18.01\ g\ mol^{-1}) + 10.0\ g/62.08\ g\ mol^{-1}}\right)$$

$$= 760\ \text{torr}\left(\frac{4.997}{4.997 + 0.161}\right) = \mathbf{736\ torr}$$

14 CHEMICAL THERMODYNAMICS

14.1 **Thermo** implies heat; **dynamics** implies movement or change.

14.3 Pressure-volume work and electrical work.

14.5 A **reversible process** is one that can be made to reverse its direction by the smallest change in an opposing force such as pressure or temperature.

14.7 The first law of thermodynamics states that if a system undergoes some series of changes that ultimately brings it back to its original state, the net energy change is zero.

14.9 A "perpetual motion machine" is a device that could run forever without a net consumption of energy. Such a machine is not possible because it would violate the first law of thermodynamics by creating energy to run the machine.

14.11 (a) There would be no change in temperature. There are no attractive forces between the molecules in an ideal gas, so there would be no change in PE when the molecules move further apart.

(b) Since gases cool on expansion, the average kinetic energy of the molecules must decrease. Therefore, ΔE is positive, q is positive, w = 0. Heat would have to be supplied to keep the temperature of the system constant (isothermal).

14.13 The extra P.E. can't just disappear. It shows up as an increase in the heat of reaction.

14.15 ΔE is the heat of reaction at constant volume, whereas ΔH is the heat of reaction at constant pressure. Most reactions that are of interest to us take place at constant P, not constant V.

14.17 $w = -P\Delta V$ $\qquad V_2 = 50.0\ m^3 \times \frac{200\ kPa}{100\ kPa} = 100\ m^3$

$w = -100\ kPa \times (100\ m^3 - 50.0\ m^3) = -5{,}000\ kPa\ m^3 = \mathbf{-5.0 \times 10^3\ kJ}$

$\mathbf{-w = q = 5.0 \times 10^3\ kJ}$

14.19 (a) **q= 35 J, w = 40 J,** $\Delta E_{system} = q + w = 35\ J + (+40J) = \mathbf{75\ J}$

(b) $\mathbf{q_{surr.} = -35\ J,\ w_{surr.} = -40\ J,\ \Delta E_{surr.} = -75\ J}$

14.21 In reaction (a) (only) the moles of gaseous material is increasing as the reaction goes from left to right. The increase in number of moles of gas will result in an increase in volume at constant pressure and temperature. Therefore, for the reaction, $P\Delta V$ will be positive and ΔE will be a larger negative value than ΔH according to the equation: $\Delta H = \Delta E + P\Delta V$.

14.23 The calculated $\Delta H_f°$ is usually from average bond energies.
Tabulated bond energies are averages as the bond energies for each kind of bond differ slightly for different compounds.

14.25 $3H_2(g) + 6C(s) \rightarrow C_6H_6(g)$

$3H_2(g) \rightarrow 6H(g)$ ΔH_1

$6C(s) \rightarrow 6C(g)$ ΔH_2

$6H(g) + 6C(g) \rightarrow C_6H_6(g)$ ΔH_3

ΔH_3 = formation of 3 (C=C), 3(C-C), and 6(H-C) bonds.

$\Delta H_f^\circ = \Delta H_1 + \Delta H_2 + \Delta H_3$
= 6(218 kJ) +6(715 kJ) + [3(-607 kJ) +3(-348 kJ) +6(-415 kJ)] = **243 kJ/mol**

Resonance energy = 243 kJ -82.8 kJ =**160 kJ**. Compounds that have resonance structures are much more stable than predicted by non-resonance structures.

14.27 $3C(s) + 4H_2(g) \rightarrow CH_3CH_2CH_3$ $\Delta H_f^\circ = ?$

$4H_2(g) \rightarrow 8H(g)$ ΔH_1

$3C(s) \rightarrow 3C(g)$ ΔH_2

$8H(g) + 3C(g) \rightarrow CH_3CH_2CH_3$ ΔH_3

ΔH_3 = formation of 8 (C-H) and 2 (C-C) bonds
$\Delta H_f^\circ = \Delta H_1 + \Delta H_2 + \Delta H_3$
= 8(218 kJ) + 3(715 kJ) + [8(-415 kJ) + 2(-348 kJ)] = **-127 kJ mol^{-1}**
Literature value is -104 kJ mol^{-1} (Table 6.1)

14.29 Many spontaneous reactions give off energy. However, there are some instances where energy is absorbed during a spontaneous reaction. Both the change in energy (heat content) and in degree of randomness (entropy) must be considered when discussing the likelihood of a reaction being spontaneous.

14.31 ΔS must be positive and the product of $T\Delta S$ must be greater than ΔH.

14.33 (a), (b), and (d)

14.35 (a) positive (b) negative (c) positive (d) positive (e) negative

14.37 During any spontaneous change, there is always an increase in the entropy of the universe.

14.39 Because of the amount of energy required to reverse the spontaneous distribution of the pollutants,i.e., must counteract the enormous entropy increase.

14.41 Between 10,000 and 20,000 atm. It is not theoretically possible to change graphite to diamond at 1 atm.

14.43 There is perfect order (zero randomness) in a pure crystalline substance at absolute zero. A mixture would have a positive entropy at 0 K because of the random distribution of particles throughout the mixture.

14.45 (a) $C(s)$ (graphite) + $2Cl_2(g) \rightarrow CCl_4(\ell)$

$\Delta S = (214.4\ J\ K^{-1}) - [5.69\ J\ K^{-1} + (2 \times 223.0)\ J\ K^{-1}]$

$\Delta S_f^\circ = \mathbf{-237.3\ J\ K^{-1}}$

(b) $Mg(s) + O_2(g) + H_2(g) \rightarrow Mg(OH)_2(s)$

$\Delta S = (63.1\ J\ K^{-1}) - [32.5\ J\ K^{-1} + 205.0\ J\ K^{-1} + 130.6\ J\ K^{-1}]$

$\Delta S_f^\circ = \mathbf{-305.0\ J\ K^{-1}}$

(c) $Pb(s) + S(s) + 2O_2(g) \rightarrow PbSO_4(s)$ $\qquad \Delta S_f^\circ = \mathbf{-358\ J\ K^{-1}}$

(d) $Na(s) + 1/2H_2(g) + C(s) + 3/2O_2(g) \rightarrow NaHCO_3(s)$

$\Delta S_f^\circ = \mathbf{-274\ J\ K^{-1}}$

(e) $1/2N_2(g) + 3/2H_2(g) \rightarrow NH_3(g)$ $\qquad \Delta S_f^\circ = \mathbf{-99.2\ J\ K^{-1}}$

14.47 (a) $\Delta S° = [S°_{Al_2O_3} + 2S°_{Fe}] - [2S°_{Al} + S°_{Fe_2O_3}]$
$= [51.0\ J/K + 2(27.3)\ J/K] - [\ 2(28.3)\ J/K + 87.4\ J/K]$
$=$ **-38.4 J/K**

(b) $\Delta S° = [S°_{SiO_2} + 2S°_{H_2O}] - [S°_{SiH_4} + 2S°_{O_2}]$
$= [41.8\ J/K + 2(188.7)\ J/K] - [205\ J/K + 2(205.0)\ J/K]$
$=$ **-196 J/K**

(c) $\Delta S° = [S°_{CaSO_4}] - [S°_{CaO} + S°_{SO_3}]$
$= (107\ J/K) - (39.8\ J/K + 256\ J/K)$
$=$ **-189 J/K**

(d) $\Delta S° = [S°_{Cu} + S°_{H_2O}] - [S°_{CuO} + S°_{H_2}]$
$= (33.15\ J/K + 188.7\ J/K) - (42.6\ J/K + 130.6)$
$=$ **48.6 J/K**

(e) $\Delta S° = [S°_{C_2H_6}] - [S_{C_2H_4} + S°_{H_2}]$
$= (230\ J/K) - (220\ J/K + 130.6\ J/K)$
$=$ **-121 J/K**

14.49 $\Delta G°$ = sum of ΔG_f's of products - sum of ΔG_f's of reactants

(a) $2Al(s) + Fe_2O_3(s) \rightarrow Al_2O_3(s) + 2Fe(s)$

$$\Delta G° = [1\ mol \times \frac{(-1577\ kJ)}{mol} + 2\ mol\ (0)] - [2\ mol\ (0) + 1\ mol \times \frac{(-741.0\ kJ)}{mol}]$$

$= -1577\ kJ + 741.0\ kJ =$ **-836 kJ**

(b) $SiH_4(g) + 2O_2(g) \rightarrow SiO_2(s) + 2H_2O(g)$
$\Delta G° = [1\ (-856\ kJ) + 2\ (-228\ kJ)] - [1\ (+52.3\ kJ) + 2(0)] =$ **-1364 kJ**

(c) $CaO(s) + SO_3(g) \rightarrow CaSO_4(s)$ $\Delta G° =$ **-346 kJ**

(d) $CuO(s) + H_2(g) \rightarrow Cu(s) + H_2O(g)$ $\Delta G° =$ **-101 kJ**

(e) $C_2H_4(g) + H_2(g) \rightarrow C_2H_6(g)$ $\Delta G° =$ **-101 kJ**

14.51 $\Delta G° = [6 \times \Delta G_f°(CO_2) + 6 \times \Delta G_f°(H_2O)] - [1 \times \Delta G_f°(glu) + 6 \times \Delta G_f°(O_2)]$
$= [6 \times (-395\ kJ) + 6 \times (-237\ kJ)] - [1 \times (-910.2\ kJ) + 6 \times (0)] =$ **-2882 kJ**
or -2880 kJ (only 3 sign. figs)

14.53 $3CaCO_3(s) \rightarrow 3\ CaO(s) + 3CO_2(g)$ $\Delta G_1 = 3(+130\ kJ)$
$3CaO(s) + 2H_3PO_4(\ell) \rightarrow Ca_3(PO_4)_2(s) + 3H_2O(\ell)$ $\Delta G_2 = 1(-512\ kJ)$

$3\ CaCO_3(s) + 2H_3PO_4(\ell) \rightarrow Ca_3(PO_4)_2(s) + 3CO_2(g) + 3H_2O(\ell)$
$\Delta G_3 = \Delta G_1 + \Delta G_2$
$= 3(130\ kJ) + (-512\ kJ)$
$\Delta G_3 =$ **-122 kJ**

14.55 Advantage - maximum work is obtained.
Disadvantage - the change takes forever to occur.

14.57 See Figure 14.14 (c) The position of the equilibrium favors products.

14.59 At equilibrium at constant pressure: $0 = \Delta H - T\Delta S$ or $\Delta S = \frac{\Delta H}{T}$

$\Delta S_{vap} = 40.7 \times 10^3\ J\ mol^{-1}/373\ K =$ **109 J mol^{-1} K^{-1}**
$\Delta S_{fus} = 6.02 \times 10^3\ J\ mol^{-1}/273\ K =$ **22.1 J mol^{-1} K^{-1}**
Both should be positive since both processes increase the randomness of the system. One would expect vaporization to have a greater increase in randomness than does melting and the above values verify that expectation.

14.61 $\Delta G°$ determines the position of equilibrium between reactants and products. Whether we start with pure reactants or pure products, some reaction will occur (accompanied by a free energy decrease) until equilibrium is reached.

14.63 There is none.

14.65 See Figure 14.14.

14.67 First calculate the moles present in 4.00 L of $C_4H_{10}(g)$ at 25°C and 1 atm.

$PV = nRT \quad n = [(1\ atm)(4.00\ L)]/[(0.0821\ L\ atm\ mol^{-1}\ K^{-1})(298\ K)]$

n = 0.163 (if 1 atm is assumed to have 3 significant figures)

Calculate ΔG° for: $C_4H_{10}(g) + 13/2O_2(g) \rightarrow 4CO_2(g) + 5H_2O(\ell)$

ΔG° = [4(-395) kJ + 5(-237)kJ] -[1(-17.0)kJ + 13/2(0)]
= -2750 kJ per mole C_4H_{10}

For 0.163 mole there would be 2750 x 0.163 or **448 kJ** of useful work.

15 CHEMICAL EQUILIBRIUM IN GASEOUS SYSTEMS

15.1 In a dynamic equilibrium, products are constantly changing to reactants and reactants are constantly changing to products but the overall effect is no net change in concentration.

15.3 (a) $\dfrac{p_{NO}^2}{p_{N_2} p_{O_2}} = K_p$ (b) $\dfrac{p_{NO_2}^2}{p_{NO}^2 p_{O_2}} = K_p$ (c) $\dfrac{p_{H_2S}^2}{p_{H_2}^2 p_{S_2}} = K_p$

(d) $\dfrac{p_{NO_2}^4 p_{O_2}}{p_{N_2O_5}^2} = K_p$ (e) $\dfrac{p_{POCl_3}^{10}}{p_{P_4O_{10}} p_{PCl_5}^6} = K_p$

15.5 By convention. This simplifies tabulation of equilibrium constants by removing ambiguity.

15.7 (a) $\frac{[HCl]^2}{[H_2][Cl_2]}$ or $\frac{p_{HCl}^2}{p_{H_2} p_{Cl_2}}$

(b) $\frac{[HCl]}{[H_2]^{1/2}[Cl_2]^{1/2}}$ or $\frac{p_{HCl}}{p_{H_2}^{1/2} p_{Cl_2}^{1/2}}$ K(a) would equal K(b) squared.

15.9 $K_p = \frac{P_{NO}^2}{P_{N_2} P_{O_2}} = \frac{P_{NO}^2 P_{H_2O}^3}{P_{NH_3}^2 P_{O_2}^{5/2}} \times \frac{P_{NH_3}^2}{P_{N_2} P_{H_2}^3} \times \frac{P_{H_2}^3 P_{O_2}^{3/2}}{P_{H_2O}^3}$

$= (9 \times 10^{172})^{1/2} \times (9.1 \times 10^5) \times (8.6 \times 10^{79})^{-3/2}$

$= \mathbf{3 \times 10^{-28}}$

15.11 By looking at the size of the equilibrium constant, you can determine whether the reaction favors the forward or reverse reaction. If the number is much greater than one, the reaction will tend to proceed far toward completion. If, however, the K is much less than one, only small amounts of products will be present at equilibrium.

15.13 Of the reactions given, reaction (a) will proceed the farthest toward completion if allowed to come to equilibrium. Reaction (b) will proceed the least toward completion.

15.15 $\Delta G° = -RT \ln(1) = \mathbf{0}$

15.17 Equation 14.7 $\Delta G = \Delta H - T\Delta S$ or $\Delta G^\circ = \Delta H^\circ - T\Delta S^\circ$
Equation 15.5 $\Delta G^\circ = -RT \ln K_P$
When the two equations are combined and ln changed to log, one obtains:

$\Delta H^\circ - T\Delta S^\circ = -2.303\ RT \log K_P$

$$\log K_p = \frac{T\Delta S^\circ - \Delta H^\circ}{2.303\ RT} = \frac{T\Delta S^\circ}{2.303\ RT} - \frac{\Delta H^\circ}{2.303\ RT}$$

$$\log K_p = \frac{\Delta S^\circ}{2.303\ R} - \frac{\Delta H^\circ}{2.303\ R} \times \frac{1}{T}$$

(This is in the form $y = b + mx$, where $\log K_p$ corresponds to y and 1/T corresponds to x.)

Slope = m = $-\Delta H^\circ/2.303\ R$ y intercept = $\Delta S^\circ/2.303\ R$
Slope gives ΔH° **y intercept gives ΔS°**

15.19 $\Delta G^\circ = -R\ (298\ K) \ln K_P$
ΔG° = sum of ΔG_f° products - sum of ΔG_f° reactants
ΔG° = [1 mol(86.8 kJ/mol) + 1 mol(-370 kJ/mol)] - [1 mol(-300 kJ /mol) + 1 mol(+51.9 kJ/mol)] = -35.1 kJ

-35.1×10^3 J (per mol) = $-(8.314\ J\ mol^{-1}K^{-1})(298\ K) \ln K_P$

$K_P = 1 \times 10^6$

15.21 $\Delta G = -RT \ln K_P$
-13.5×10^3 J (per mol) = $-(8.314\ J\ mol^{-1}\ K^{-1})(700\ K) \ln K_P$

$2.32 = \ln K_P$ $K_P =$ **10**

15.23 527°C = 800 K $\Delta G = -RT \ln K_P$
$\Delta G = -(8.314\ J\ mol^{-1}\ K^{-1})(800\ K) \ln 5.10$

$\Delta G = -1.08 \times 10^4$ J = **-11 kJ**

15.25 $\Delta H° = [1 \text{ mol } (-84.5 \text{ kJ mol}^{-1})] - [1 \text{ mol } (51.9 \text{ kJ mol}^{-1}) + 1(0)] = -136.4 \text{ kJ}$

$\Delta S° = [1 \text{ mol } (230 \text{ J mol}^{-1} \text{ K}^{-1})] - [1 \text{ mol } (220 \text{ J mol}^{-1} \text{ K}^{-1}) + 1 \text{ mol } (130.6 \text{ J mol}^{-1} \text{ K}^{-1})]$
$= -121 \text{ J K}^{-1}$

$\Delta G = -RT \ln K_P = \Delta H - T\Delta S$

$\Delta H = T\Delta S - RT \ln K_P = T(\Delta S - R \ln K_P)$

$$\frac{\Delta H}{\Delta S - R \ln K_P} = T$$

$$\frac{-136.4 \times 10^3 \text{ J}}{(-120.6 \text{ J K}^{-1}) - (8.314 \text{ J mol}^{-1} \text{ K}^{-1})(\ln 1)} = T$$

$$T = \frac{-136.4 \times 10^3 \text{K}}{120.6 + 0} = \mathbf{1.13 \times 10^3 \text{ K}} \text{ or } 860°$$

15.27 15.2 (a) and 15.4 (b), only

15.29 $K_P = K_c (RT)^{\Delta n}$
$K_P = (5.67 \text{ mol}^2/\text{L}^2)[(0.0821 \text{ L atm mol}^{-1} \text{ K}^{-1})(1773 \text{ K})]^2$
$\mathbf{K_P} = (5.67)(2.119 \times 10^4) \text{ atm}^2 = \mathbf{1.20 \times 10^5 \text{ atm}^2}$

15.31 At equilibrium, $K_c = [H_2O(g)]$, $K_p = p_{H_2O(g)}$

Thus, p_{H_2O} or $[H_2O]$ are constants that only change with temperature.

15.33 (a) $K_c = [CO_2(g)]$

(b) $K_c = \frac{[Ni(CO)_4(g)]}{[CO(g)]^4}$

(c) $K_c = \frac{[I_2(g)][CO_2(g)]^5}{[CO(g)]^5}$

(d) $K_c = \frac{[CO_2(g)]}{[Ca(HCO_3)_2(aq)]}$

(e) $K_c = [Ag^+(aq)][Cl^-(aq)]$

15.35 None of the above will effect the equilibrium constant for the reaction. The only change which will effect the equilibrium constant is a change in temperature.

15.37 (a) **No change in the value of K** (b) **No change in the value of K**

(c) **No change in the value of K** (d) **The value of K will increase**

(e) **No change in the value of K**

15.39 (a) **decreased** (b) **increased** (c) **no change** (d) **decreased**

15.41 (a) **no change** (b) **increased** (c) **decreased** (d) **increased**

15.43 $K_c = 5.5 \quad \Delta n = +1$
$K_P = K_c(RT)^{\Delta n} = (5.5 \text{ mol/L})[(0.0821 \text{ L atm mol}^{-1} \text{ K}^{-1})(298 \text{ K})]^1$
$K_P = (5.5)(2.447 \times 10^{+1} \text{ atm}) = \mathbf{1.3 \times 10^2 \text{ atm}}$

15.45 $K_c = \frac{[PCl_3]\,[Cl_2]}{[PCl_5]} = \frac{[PCl_3]\,(1.87 \times 10^{-1})}{1.29 \times 10^{-3}} = 33.3$

$[PCl_3] = \mathbf{2.30 \times 10^{-1}\ mol\ L^{-1}}$

15.47 $K_c = \frac{[H_2]\,[I_2]}{[HI]^2} = \frac{(1.0 \times 10^{-3}\ M)(2.5 \times 10^{-2}\ M)}{(2.2 \times 10^{-2}\ M)^2} = \mathbf{5.2 \times 10^{-2}}$

15.49

	$2N_2O$ +	$3O_2$ ⇌	$4NO_2$
Init. Conc.	0.020	0.0560	0
Change	- 2X	-3X	+4X
Equil. Conc.	0.020 - 2X	0.0560 - 3X	0.020 or 4X

(a) $4X = 0.020 \qquad X = 5.0 \times 10^{-3}$

$[N_2O] = 0.020 - 2X = 0.020 - 2(5.0 \times 10^{-3}) = \mathbf{0.010\ M}$

$[O_2] = 0.0560 - 3X = 0.0560 - 3(5.0 \times 10^{-3}) = \mathbf{0.041\ M}$

(b) $K_c = \frac{[NO_2]^4}{[N_2O]^2[O_2]^3} = \frac{(0.020)^4}{(0.010)^2(0.041)^3} = \mathbf{23\ L\ mol^{-1}}$

15.51

	H_2 +	CO_2 ⇌	CO +	H_2O
Init. Conc.	0.200	0.200	0	0
Change	-X	-X	+X	+X
Equil. Conc.	0.200 -X	0.200 - X	+X	+X

$$K_c = \frac{(X)^2}{(0.200 - X)^2} = 0.771$$

$$\sqrt{0.771} = \frac{X}{(0.200 - X)}$$

X = 0.0935

$[H_2]$ = $[CO_2]$ = 0.106 M **[CO] = $[H_2O]$ = 0.0935 M**

15.53 (a)

	$2CO_2$ ⇌	$2CO$ +	O_2
Init. Conc.	1.0×10^{-3} M	0	0
Change	-2X	+2X	+X
Equil. Conc.	1.0×10^{-3} -2X	2X	X

$$K_c = 6.4 \times 10^{-7} = \frac{(2X)^2 X}{(1.0 \times 10^{-3} - 2X)^2}$$

If 2X is small compared to 1.0×10^{-3}, then:

$$6.4 \times 10^{-7} = \frac{(2X)^2 X}{(1.0 \times 10^{-3})^2} = \frac{4X^3}{(1.0 \times 10^{-3})^2}$$

$X = 5.4 \times 10^{-5}$ Check! Was 2X small compared to 1.0×10^{-3}? The value obtained for 2X was about 11% of the1.0×10^{-3}. This is about the limit allowed in most approximations. Using this value, the equilibrium concentrations are:

$[CO_2]$ = 8.9×10^{-4} M **[CO] = 1.1×10^{-4} M** **$[O_2]$ = 5.4×10^{-5} M**

(continued)

15.53 (continued)

To obtain a more precise solution, one would solve the above equation using a series of approximations. The above would be the first approximation. These values would be substituted into the equilibrium expression and the process would be repeated. The values after successive approximations are:

$[CO_2]$ = 9.0 x 10^{-4} M,

[CO] = 1.0 x 10^{-4} M,

$[O_2]$ = 5.1 x 10^{-5} M

(b) $\dfrac{(1.0 \times 10^{-3} - 9.0 \times 10^{-4})}{1.0 \times 10^{-3}} = \mathbf{0.10}$ = the fraction of CO_2 decomposed

15.55 $K_P = P_{CO_2} \times P_{H_2O} = 0.25 \text{ atm}^2 = X^2$

$$P_{CO_2} = \sqrt{0.25 \text{ atm}^2} = \mathbf{0.50 \text{ atm}} = \mathbf{P_{CO_2} = P_{H_2O}}$$

It is used in baking because it liberates gaseous CO_2 and H_2O that are trapped in the dough, thus causing it to rise.

15.57

	$2NO_2 \rightleftharpoons$	N_2O_4
Init. Conc.	1.0 M	0
Change	- 2X	+X
Equil. Conc.	1.0 - 2X	X

$$7.5 = \frac{X}{(1.0 - 2X)^2}$$

$30X^2 - 31X + 7.5 = 0$

(continued)

15.57 (continued)

X = 0.3865 (the logical solution of the two solutions of the quadratic equation.)

$[NO_2]$ = 1.0 - 2 x 0.3865 = 0.23 M

$[N_2O_4]$ = 0.39 M

Double the size of the container!

	$2NO_2$ $\rightleftharpoons$	N_2O_4
Init. Conc.	0.50 M	0
Change	-2X	X
Equil. Conc.	0.50 - 2X	X

$$7.5 = \frac{X}{(0.50 - 2X)^2}$$

$30X^2 - 16X + 1.875 = 0$ X = 0.174

$[NO_2]$ = 0.50 - 2 x 0.174 = 0.15 M

$[N_2O_4]$ = 0.17 M

Yes! The larger container favors the NO_2 while the smaller container favored the N_2O_4. This is what one should expect based upon a knowledge of LeChatelier's Principle.

15.59

	Initial Concentration	Change	Equilibrium Concentration
H_2	0.0200	- X	0.0200 - X
CO_2	0.0400	- X	0.0400 -X
CO	0	+X	X
H_2O	0	+X	X

$$K_c = 0.771 = \frac{(X)(X)}{(0.0200 - X)(0.0400 - X)} = \frac{X^2}{(8.00 \times 10^{-4}) - (0.0600\ X) + X^2}$$

$6.17 \times 10^{-4} - 4.63 \times 10^{-2}\,X + 0.771\ X^2 = X^2$

$.229\ X^2 + (4.63 \times 10^{-2})X - 6.17 \times 10^{-4} = 0$

$X = 1.25 \times 10^{-2}$

$[H_2] = 7.5 \times 10^{-3}\ M$ $[CO_2] = 2.75 \times 10^{-2}\ M$

$[CO] = 1.25 \times 10^{-2}\ M$ $[H_2O] = 1.25 \times 10^{-2}\ M$

15.61

	N_2	+	O_2	$\rightleftharpoons$	2NO
Init. p	33.6		4.0		0
Change	-X		- X		+ 2X
Equil. p	33.6 - X		4.0 - X		2X

$$K_P = 4.8 \times 10^{-7} = \frac{(2X)^2}{(33.6 - X)(4.0 - X)} \approx \frac{(2X)^2}{(33.6)(4.0)}$$

$X = 4.0 \times 10^{-3}$

$p_{N_2} = 33.6$ atm $p_{O_2} = 4.0$ atm $p_{NO} = 8.0 \times 10^{-3}$ atm

16 ACID-BASE EQUILIBRIA IN AQUEOUS SOLUTIONS

16.1 $H_2O + H_2O \rightleftharpoons H_3O^+(aq) + OH^-(aq)$
"This is a very important equilibrium because it is present in any aqueous solution, regardless of what other reactions may also be taking place."

16.3 $HCl(g) + H_2O(\ell) \rightarrow H^+(aq) + Cl^-(aq)$
or $\rightarrow H_3O^+(aq) + Cl^-(aq)$
$KOH(s) + H_2O(\ell) \rightarrow K^+(aq) + OH^-(aq)$

16.5 (a) $[H^+] = 1.0 \times 10^{-3}$ mol/L $[OH^-] = 1.0 \times 10^{-11}$ mol/L pH = 3.00

(b) $[H^+] = 1.25 \times 10^{-1}$ mol/L $[OH^-] = 8.00 \times 10^{-14}$ mol/L pH = 0.903

(c) $[H^+] = 3.2 \times 10^{-12}$ mol/L $[OH^-] = 3.1 \times 10^{-3}$ mol/L pH = 11.49

(d) $[H^+] = 4.2 \times 10^{-13}$ mol/L $[OH^-] = 2.4 \times 10^{-2}$ mol/L pH = 12.38

(e) $[H^+] = 2.1 \times 10^{-4}$ mol/L $[OH^-] = 4.8 \times 10^{-11}$ mol/L pH = 3.68

(continued)

16.5 (continued)

(f) $[H^+] = \mathbf{1.3 \times 10^{-5}}$ **mol/L** $[OH^-] = \mathbf{7.7 \times 10^{-10}}$ **mol/L** pH = **4.89**

(g) $[H^+] = \mathbf{1.2 \times 10^{-12}}$ **mol/L** $[OH^-] = \mathbf{8.4 \times 10^{-3}}$ **mol/L** pH = **11.92**

(h) $[H^+] = \mathbf{2.1 \times 10^{-13}}$ **mol/L** $[OH^-] = \mathbf{4.8 \times 10^{-2}}$ **mol/L** pH = **12.68**

16.7 $pH = -\log [H^+]$; $pOH = -\log [OH^-]$;
$K_w = [H^+][OH^-]$ $\log K_w = \log [H^+] + \log [OH^-]$
$-\log K_w = -\log (1 \times 10^{-14}) = -\log [H^+] - \log [OH^-]$ $pK_w = 14 = pH + pOH$

16.9 **(e)<(b)<(c)<(d)<(a)**

16.11 (a) $[H^+] = \mathbf{0.050}$ **mol/L** $[OH^-] = \mathbf{2.0 \times 10^{-13}}$ **mol/L**

(b) $[H^+] = \mathbf{1.9 \times 10^{-6}}$ **mol/L** $[OH^-] = \mathbf{5.3 \times 10^{-9}}$ **mol/L**

(c) $[H^+] = \mathbf{1.0 \times 10^{-4}}$ **mol/L** $[OH^-] = \mathbf{1.0 \times 10^{-10}}$ **mol/L**

(d) $[H^+] = \mathbf{1.6 \times 10^{-8}}$ **mol/L** $[OH^-] = \mathbf{6.2 \times 10^{-7}}$ **mol/L**

(e) $[H^+] = \mathbf{1.1 \times 10^{-11}}$ **mol/L** $[OH^-] = \mathbf{9.1 \times 10^{-4}}$ **mol/L**

(f) $[H^+] = \mathbf{2.5 \times 10^{-13}}$ **mol/L** $[OH^-] = \mathbf{4.0 \times 10^{-2}}$ **mol/L**

16.13 $pH = -\log [H^+]$ $pOH = -\log[OH^-]$

(a) pH = **3.00** pOH = **11.00**

(b) pH = **0.903** pOH = **13.097**

(c) pH = **11.49** pOH = **2.51**

(d) pH = **12.38** pOH = **1.62**

(continued)

16.13 (continued)

(e) pH = **3.68** pOH = **10.32**

(f) pH = **4.89** pOH = **9.11**

(g) pH = **11.92** pOH = **2.08**

(h) pH = **12.68** pOH = **1.32**

16.15 (a) $C_5H_5N + H_2O \rightleftharpoons C_5H_5NH^+ + OH^-$

(b) $CO_3^{2-} + H_2O \rightleftharpoons HCO_3^- + OH^-$

(c) $H_2PO_4^- + H_2O \rightleftharpoons H_3PO_4 + OH^-$

(d) $NO_2^- + H_2O \rightleftharpoons HNO_2 + OH^-$

(e) $C_6H_5NH_2 + H_2O \rightleftharpoons C_6H_5NH_3^+ + OH^-$

16.17 (d) < (c) < (a) < (b)

16.19 PH_2^- is a stronger base than HS^-.

16.21 Ammonia

16.23 $pK_a = -\log K_a = -\log 3.8 \times 10^{-9} =$ **8.42**

$K_b = \text{antilog}(-pK_b) = \text{antilog}(-3.84) =$ **1.4 x 10^{-4}**

16.25 (a) $HC_7H_5O_2 \rightleftharpoons H^+ + C_7H_5O_2^-$

$K_a = \{[H^+][C_7H_5O_2^-]\}/[HC_7H_5O_2]$

(b) $N_2H_4 + H_2O \rightleftharpoons N_2H_5^+ + OH^-$

$K_b = \{[N_2H_5^+][OH^-]\}/[N_2H_4]$

(c) $HCHO_2 \rightleftharpoons H^+ + CHO_2^-$

$K_a = \{[H^+][CHO_2^-]\}/[HCHO_2]$

(d) $HC_8H_{11}N_2O_3 \rightleftharpoons H^+ + C_8H_{11}N_2O_3^-$

$K_a = \{[H^+][C_8H_{11}N_2O_3^-]\}/[HC_8H_{11}N_2O_3]$

(e) $C_5H_5N + H_2O \rightleftharpoons C_5H_5NH^+ + OH^-$

$K_b = \{[C_5H_5NH^+][OH^-]\}/[C_5H_5N]$

16.27 (a) $NH_3 + H_2O \rightleftharpoons NH_4^+ + OH^-$ $K_b = 1.8 \times 10^{-5}$

	Init. Conc.	Change	Equil. Conc.
NH_4^+	0	+X	X
OH^-	~0	+X	X
NH_3	0.15 M	-X	$0.15 - X \approx 0.15$

$$1.8 \times 10^{-5} = \frac{(X)(X)}{(0.15 - X)} \approx \frac{X^2}{0.15}$$ **X = [OH⁻] = 1.6 x 10⁻³ mol/L**

(continued)

16.27 (continued)

(b) $N_2H_4 + H_2O \rightleftharpoons N_2H_5^+ + OH^-$ $K_b = 1.7 \times 10^{-6}$

	Init. Conc.	Change	Equil. Conc.
$N_2H_5^+$	0	+X	X
OH^-	~0	+X	X
N_2H_4	0.20	-X	$0.20 - X \approx 0.20$

$$1.7 \times 10^{-6} = \frac{(X)(X)}{(0.20 - X)} \approx \frac{X^2}{0.20}$$ $X = [OH^-] = \mathbf{5.8 \times 10^{-4}}$ **mol/L**

(c) $K_b = 3.7 \times 10^{-4} = \dfrac{X^2}{0.80 - X}$ $X = [OH^-] = \mathbf{1.7 \times 10^{-2}}$ **mol/L**

(d) $K_b = 1.1 \times 10^{-8} = \dfrac{X^2}{0.35 - X}$ $X = [OH^-] = \mathbf{6.2 \times 10^{-5}}$ **mol/L**

(e) $K_b = 1.7 \times 10^{-9} = \dfrac{X^2}{0.010 - X}$ $X = [OH^-] = \mathbf{4.1 \times 10^{-6}}$ **mol/L**

16.29 $[H^+][OH^-] = 1 \times 10^{-14}$, $pH + (-\log [OH^-]) = 14$, $pH = 14 + \log [OH^-]$

(a) $pH = 14 + \log (1.6 \times 10^{-3}) = 14 - 2.8 =$ **11.20**

(b) pH = **10.76** (c) pH = **12.23** (d) pH = **9.79** (e) pH = **8.61**

16.31 $K_a = \frac{[H^+][\text{anion}]}{[\text{undissoc. Acid}]} = \frac{(X)(X)}{(0.10 - X)}$

pH = 5.37 = - log[H+] = - log (X)　　　　X = 4.3 x 10^{-6}

$$K_a = \frac{(4.3 \times 10^{-6})^2}{0.10 - (4.3 \times 10^{-6})} = \frac{1.8 \times 10^{-11}}{0.10} = \mathbf{1.8 \times 10^{-10}}$$

16.33 (a) $HCHO_2 \rightleftharpoons H^+ + CHO_2^-$　　K_a = 1.8 x 10^{-4}

	Init. Conc.	Change	Equil. Conc.
H^+	~0	+X	X
CHO_2^-	0	+X	X
$HCHO_2$	1.0	-X	1.0 - X

$$K_a = 1.8 \times 10^{-4} = \frac{(X)(X)}{1.0 - X} \approx \frac{X^2}{1.0}$$

X = [H+] = 1.3 x 10^{-2} M　　　% ionization = $\frac{1.3 \times 10^{-2}}{1.0}$ x 100% = **1.3 %**

(b) $K_a = 1.4 \times 10^{-5} = \frac{X^2}{0.010 - X}$　　　X = [H+] = $[C_3H_5O_2^-]$

X = 3.7 x 10^{-4}　　　% ionization = $\frac{3.7 \times 10^{-4}}{0.010}$ x 100% = **3.7 %**

(continued)

16.33 (continued)

(c) $K_a = 4.9 \times 10^{-10} = \dfrac{X^2}{0.025 - X}$ $X = 3.5 \times 10^{-6}$

$$\% \text{ ionization} = \frac{3.5 \times 10^{-6}}{0.025} \times 100\% = \mathbf{0.014\%}$$

(d) $K_a = 1.4 \times 10^{-5} = \dfrac{X^2}{0.35 - X}$ $X = 2.2 \times 10^{-3}$

$$\% \text{ ionization} = \frac{2.2 \times 10^{-3}}{0.35} \times 100\% = \mathbf{0.63\%}$$

(e) $K_a = 3.1 \times 10^{-8} = \dfrac{X^2}{0.50 - X}$ $X = 1.2 \times 10^{-4}$

$$\% \text{ ionization} = \frac{1.2 \times 10^{-4}}{0.50} \times 100\% = \mathbf{0.024\%}$$

(f) strong acid, assume **100% ionization**

16.35 $B + H_2O \rightleftharpoons HB^+ + OH^-$

$$K_b = \frac{[HB^+][OH^-]}{[B]}$$

From the pH, $[OH^-] = 2.5 \times 10^{-3}$
$[HB^+] = [OH^-] = 2.5 \times 10^{-3}$
$[B] = 0.012 - (2.5 \times 10^{-3})$

$$K_b = \frac{(2.5 \times 10^{-3})(2.5 \times 10^{-3})}{0.012 - (2.5 \times 10^{-3})} = \mathbf{6.6 \times 10^{-4}}$$

16.37 $HC_4H_3N_2O_3 \rightleftharpoons H^+ + C_4H_3N_2O_3^-$

$K_a = 1.0 \times 10^{-5}$, $M (NaC_4H_3N_2O_3)$ = init. conc. $C_4H_3N_2O_3^-$ =

$$10 \text{ mg} \times \frac{1 \text{ g}}{1000 \text{ mg}} \times \frac{1 \text{ mol}}{150 \text{ g}} \times \frac{1}{0.250 \text{ L}} = 2.67 \times 10^{-4} \text{ M}$$

	Init. Conc.	Change	Equil. Conc.
H^+	0.10	$-(2.67 \times 10^{-4} - X)$	~0.10
$C_4H_3N_2O_3^-$	2.67×10^{-4}	$-(2.67 \times 10^{-4} - X)$	X
$HC_4H_3N_2O_3$	0	$(2.67 \times 10^{-4} - X)$	$(2.67 \times 10^{-4} - X)$

$$1.0 \times 10^{-5} = \frac{(0.10)(X)}{(2.67 \times 10^{-4} - X)} \approx \frac{(0.10)(X)}{(2.67 \times 10^{-4})} \qquad X = 2.67 \times 10^{-8}$$

This shows that essentially 100% of the $C_4H_3N_2O_3^-$ is converted to barbituric acid.

16.39 $K_a = 1.8 \times 10^{-5} = \dfrac{(X)(X)}{(Y - X)}$ $\qquad X = [H^+] = 3.2 \times 10^{-3}$ M (from pH)

$$1.8 \times 10^{-5} = \frac{(3.2 \times 10^{-3})^2}{Y - (3.2 \times 10^{-3})} \qquad Y = [HC_2H_3O_2] = \mathbf{0.57\ M}$$

16.41 $K_a = \dfrac{(X)(X)}{0.010 - X}$ $\qquad X = [H^+] = 2.8 \times 10^{-5}$ M (from pH)

$$K_a = \frac{(2.8 \times 10^{-5})^2}{0.010 - (2.8 \times 10^{-5})} = \mathbf{7.8 \times 10^{-8}}$$

16.43 $HCO_2H \rightleftharpoons CHO_2^- + H^+$ $K_a = 1.8 \times 10^{-4}$

	Init. Conc.	Change	Equil. Conc.
H^+	~0	+X	X
CHO_2^-	0	+X	X
$HCHO_2$	0.010	-X	0.010 - X

$$1.8 \times 10^{-4} = \frac{(X)(X)}{0.010 - X}$$

Solving this quadratic equation yields, $X = 1.3 \times 10^{-3}$ (To two significant figures, the same answer will be obtained if one assumed that $0.010\text{-}X \approx 0.010$.)

$[H^+] = 1.3 \times 10^{-3}$ M **$[CHO_2^-] = 1.3 \times 10^{-3}$ M**

$[HCHO_2] = 0.009$ M **$[OH^-] = 7.7 \times 10^{-12}$ M**

16.45 No. HCl is a strong acid and is completely dissociated. Cl^- is such a weak conjugate base that it cannot neutralize acids.

16.47 (a) $HC_2H_3O_2 \rightleftharpoons H^+ + C_2H_3O_2^-$

	Init. Conc.	Change	Equil. Conc.
H^+	~0	+X	X
$C_2H_3O_2^-$	0.15 M	+X	$0.15 + X \approx 0.15$
$HC_2H_3O_2$	0.25	-X	$0.25 - X \approx 0.25$

$$K_a = 1.8 \times 10^{-5} = \frac{(X)(0.15)}{(0.25)}$$ **$X = [H^+] = 3.0 \times 10^{-5}$ mol/L**

(b) $$1.8 \times 10^{-4} = \frac{(X)(0.50)}{(0.50)}$$ **$X = [H^+] = 1.8 \times 10^{-4}$ mol/L**

(continued)

16.47 (continued)

(c) $4.5 \times 10^{-4} = \frac{(X)(0.40)}{(0.30)}$

$X = [H^+] = \mathbf{3.4 \times 10^{-4}}$ **mol/L**

(d) $1.8 \times 10^{-5} = \frac{(X)(0.15)}{(0.25)}$ $X = [OH^-] = 3.0 \times 10^{-5}$ mol/L

$[H^+] = 1.0 \times 10^{-14} / 3.0 \times 10^{-5} = \mathbf{3.3 \times 10^{-10}}$ **mol/L**

(e) $1.7 \times 10^{-6} = \frac{(X)(0.50)}{(0.30)}$ $X = [OH^-] = 1.0 \times 10^{-6}$ mol/L

$[H^+] = 1.0 \times 10^{-14}/1.0 \times 10^{-6} = \mathbf{1.0 \times 10^{-8}}$ **mol/L**

16.49 (a) $NH_3 + H_2O \rightleftharpoons NH_4 + OH^-$

$K_b = 1.8 \times 10^{-5} = \frac{[NH_4^+][OH^-]}{[NH_3]} = \frac{(0.10)[OH^-]}{(0.10)}$ $[OH^-] = 1.8 \times 10^{-5}$

$[H^+] = 1.0 \times 10^{-14}/1.8 \times 10^{-5} = 5.6 \times 10^{-10}$ **pH = 9.26**

(b) $HC_2H_3O_2 \rightleftharpoons H^+ + C_2H_3O_2^-$ $K_a = 1.8 \times 10^{-5} = \frac{[H^+](0.40)}{(0.20)}$

$[H^+] = 9.0 \times 10^{-6}$ M **pH = 5.05**

(c) $N_2H_4 + H_2O \rightleftharpoons N_2H_5^+ + OH^-$ $K_b = 1.7 \times 10^{-6} = \frac{(0.10)[OH^-]}{(0.15)}$

$[OH^-] = 2.6 \times 10^{-6}$ **pH = 8.41**

(d) $HCl \rightarrow H^+ + Cl^-$ (not a buffer; 100% ionization) $[H^+] = 0.20$ **pH = 0.70**

16.51 pH = 10.00, $[H^+] = 1.0 \times 10^{-10}$ $[OH^-] = 1.0 \times 10^{-4}$

$NH_3 + H_2O \rightleftharpoons NH_4^+ + OH^-$

$$K_b = 1.8 \times 10^{-5} = \frac{[NH_4^+](1.0 \times 10^{-4})}{[NH_3]}$$

$$\frac{[NH_4^+]}{[NH_3]} = 0.18 \qquad \frac{[NH_3]}{[NH_4^+]} = \frac{1}{0.18} = \mathbf{5.6}$$

16.53 (a) Initial pH $1.8 \times 10^{-5} = \frac{[H^+][1.00]}{[1.00]}$ $[H^+] = 1.8 \times 10^{-5}$

pH = 4.74

pH after the addition of 0.10 mol NaOH per 0.500 L or 0.20 M; pH = ? (Note: There will not be dilution during the addition of NaOH. The 0.20 M has been calculated using the volume of the buffer.)

	Init. Conc.	Effect of NaOH	Change	Equil. Conc.
$C_2H_3O_2^-$	1.00	+0.20	+X	$1.20 + X \approx 1.20$
H^+	~0		+X	X
$HC_2H_3O_2$	1.00	-0.20	-X	$0.80 - X \approx 0.80$

$$K_a = 1.8 \times 10^{-5} = \frac{(X)(1.20)}{(0.80)}$$

$X = [H^+] = 1.2 \times 10^{-5}$

pH = 4.92 **ΔpH** = 4.92 - 4.74 = **0.18**

(b) Initial pH = 4.74 pH after the addition of 0.20 M NaOH; equals ?

	Init. Conc.	Effect of NaOH	Change	Equil. Conc.
$C_2H_3O_2^-$	0.50	+0.20	+X	$0.70 + X \approx 0.70$
H^+	~0		+X	X
$HC_2H_3O_2$	0.50	-0.20	-X	$0.30 - X \approx 0.30$

(continued)

16.53 (continued)

$K_a = 1.8 \times 10^{-5} = \frac{(X)(0.70)}{(0.30)}$ $X = [H^+] = 7.7 \times 10^{-6}$

pH = 5.11 **ΔpH** = 5.11 - 4.74 = **0.37**

(c) Initial pH = 5.11 (Same as final pH in Question 16.53 (b))
pH after the addition of the NaOH = ?

$K_a = 1.8 \times 10^{-5} = \frac{(H^+)(0.90)}{(0.10)}$ $-\log [H^+] = 5.70$

ΔpH = 0.59

(d) Initial pH = 5.35

pH after the addition of 0.20 M NaOH = ?

	Init. Conc.	Effect of NaOH	Change	Equil. Conc.
$C_2H_3O_2^-$	0.80	+0.20	-X	1.0 - X
H^+	~0		-?	?
$HC_2H_3O_2$	0.20	-0.20	+X	0 + X

Since all of the $HC_2H_3O_2$ is consumed by the NaOH, the equilibrium must shift in the direction that tends to restore $HC_2H_3O_2$. This is hydrolysis of the $C_2H_3O_2^-$.

$$C_2H_3O_2^- + H_2O \rightarrow HC_2H_3O_2 + OH^-$$

$$K_{hy} = K_b (C_2H_3O_2^-) = \frac{K_w}{K_a} = \frac{[HC_2H_3O_2][OH^-]}{[C_2H_3O_2^-]}$$

	Init. Conc.	Change by Hydrolysis	Equil. Conc.
$HC_2H_3O_2$	0	+X	X
OH^-	~0	+X	X
$C_2H_3O_2^-$	1.0	-X	1.0 - X ≈ 1.0

$\frac{1.0 \times 10^{-14}}{1.8 \times 10^{-5}} = \frac{(X)(X)}{1.0}$ $X = [OH^-] = 2.4 \times 10^{-5}$

pH = 9.38 **ΔpH** = 9.38 - 5.35 = **4.03**

(continued)

16.53 (continued)

(e) Initial pH = 5.70 (Same as final pH in part c)

After addition of NaOH, there will be: 1.00 M $C_2H_3O_2^-$, 0.00 M $HC_2H_3O_2$, and 0.10 M excess NaOH. The M of OH^- contributed by hydrolysis can be neglected when excess base is present. We will demonstrate below that its contribution is too small to change the pH.

	Init. Conc.	Change	Equil. Conc.
$C_2H_3O_2^-$	1.00	-X	$1.00 - X \approx 1.00$
OH^-	0.10	+X	$0.10 + X$
$HC_2H_3O_2$	0	+X	X

$$K_{hy} = \frac{K_w}{K_a} = 5.56 \times 10^{-10} = \frac{(X)(0.10 + X)}{1.00}$$

$X = 5.6 \times 10^{-9}$ $[OH^-] = 0.10 + 5.6 \times 10^{-9} = 0.10$

pOH = 1.00 pH = 13.00

ΔpH = 7.30

16.55 Initial pH = 3.82 Final pH = ?

	Init. Conc.	Effect of NaOH	Change	Equil. Conc.
$HCHO_2$	0.45 M	-0.20	-X	$0.25 - X \approx 0.25$
H^+	~0		+X	X
CHO_2^-	0.55 M	+0.20	+X	$0.75 + X \approx 0.75$

$$K_a = 1.8 \times 10^{-4} = \frac{(X)(0.75)}{0.25}$$

$X = [H^+] = 6.0 \times 10^{-5}$

Final pH = 4.22

ΔpH = 0.40

16.57 (a) **neutral (salt of a strong acid and a strong base)**
(b) **acidic (salt of a strong acid and a weak base)**
(c) **basic (salt of a strong base and a weak acid)**
(d) **acidic (salt of a strong acid and a weak base)**

16.59 (a) $C_2H_3O_2^- + H_2O \rightleftharpoons HC_2H_3O_2 + OH^-$

$$K_b = \frac{K_w}{K_a} = \frac{1.0 \times 10^{-14}}{1.8 \times 10^{-5}} = 5.6 \times 10^{-10}$$

$$K_b = \frac{[HC_2H_3O_2][OH^-]}{[C_2H_3O_2^-]} = \frac{(X)(X)}{(1.0 \times 10^{-3}) - X}$$

$X = 7.5 \times 10^{-7} = [OH^-]$ pOH = 6.12 **pH = 7.88**

(b) $NH_4^+ \rightarrow NH_3 + H^+$

$$K_a = \frac{K_w}{K_b} = 5.6 \times 10^{-10} = \frac{(X)(X)}{0.125 - X}$$

$X = [H^+] = 8.3 \times 10^{-6}$ **pH = 5.08**

(c) $CHO_2^- + H_2O \rightleftharpoons HCHO_2 + OH^-$

$$K_b = \frac{K_w}{K_a} = 5.6 \times 10^{-11} = \frac{(X)(X)}{0.10 - X}$$

$X = [OH^-] = 2.4 \times 10^{-6}$ pOH = 5.62 **pH = 8.38**

(d) $CN^- + H_2O \rightleftharpoons HCN + OH^-$

$$K_b = \frac{K_w}{K_a} = 2.0 \times 10^{-5} = \frac{(X)(X)}{0.10 - X}$$

$X = [OH^-] = 1.4 \times 10^{-3}$ **pH = 11.15**

(continued)

16.59 (continued)

(e) $NH_3OH^+ \rightleftharpoons NH_2OH + H^+$

$$K_a = \frac{K_w}{K_b} = 9.1 \times 10^{-7} = \frac{(X)(X)}{0.20 - X}$$

$X = [H^+] = 4.3 \times 10^{-4}$ **pH = 3.37**

16.61 $Base^- + H_2O \rightleftharpoons H\text{-}Base + OH^-$ From pH: $[OH^-] = 2.2 \times 10^{-5}$

$$K_b = \frac{K_w}{K_a} = \frac{(2.2 \times 10^{-5})(2.2 \times 10^{-5})}{0.10 - (2.2 \times 10^{-5})} = 4.8 \times 10^{-9}$$

$$\frac{K_w}{K_a} = \frac{1.0 \times 10^{-14}}{K_a} = 4.8 \times 10^{-9}$$

$\mathbf{K_a = 2.1 \times 10^{-6}}$

16.63 $C_8H_{11}N_2O_3^- + H_2O \rightleftharpoons HC_8H_{11}N_2O_3 + OH^-$

$$K_b = \frac{K_w}{K_a} = \frac{1.0 \times 10^{-14}}{3.7 \times 10^{-8}} = 2.7 \times 10^{-7}$$

$$\frac{10\ mg}{250\ mL} \times \frac{1{,}000\ mL}{L} \times \frac{1\ g}{1{,}000\ mg} \times \frac{1\ mol}{206.2\ g} = 1.9 \times 10^{-4}\ M$$

$$K_b = 2.7 \times 10^{-7} = \frac{(X)(X)}{1.9 \times 10^{-4} - X} \approx \frac{X^2}{1.9 \times 10^{-4}}$$

$X = [OH^-] = 7.2 \times 10^{-6}$ **pH = 8.86**

16.65 $C_7H_5O_2^- + H_2O \rightleftharpoons HC_7H_5O_2 + OH^-$

$$K_b = \frac{K_w}{K_a} = \frac{1.0 \times 10^{-14}}{6.5 \times 10^{-5}} = 1.5 \times 10^{-10}$$

$$K_b = \frac{[HC_7H_5O_2][OH^-]}{[C_7H_5O_2^-]} = \frac{(X)(X)}{(0.30 - X)} = 1.5 \times 10^{-10}$$

$$\frac{X^2}{0.30} \approx 1.5 \times 10^{-10}$$

$X = 6.7 \times 10^{-6}$ $[OH^-] = 6.7 \times 10^{-6}$ pOH = 5.17 **pH = 8.83**

16.67 $CN^- + H_2O \rightleftharpoons HCN + OH^-$

$$K_b = \frac{K_w}{K_a} = \frac{1.0 \times 10^{-14}}{4.9 \times 10^{-10}} = 2.0 \times 10^{-5}$$

$$K_b = 2.0 \times 10^{-5} = \frac{(X)(X)}{0.0010 - X} \approx \frac{X^2}{0.0010}$$

$X = [OH^-] = 1.4 \times 10^{-4}$

Again, we see that the variable in the denominator is not negligible. Therefore, we may not drop the "X" value in the denominator. The following solution is by successive approximation; it can also be solved using the quadratic formula.

$$2.0 \times 10^{-5} = \frac{X^2}{0.0010 - X} \text{ or } \frac{X^2}{0.0010 - 1.4 \times 10^{-4}}$$

$X = [OH^-] = 1.3 \times 10^{-4}$

pH = 10.11

16.69 $H_3C_6H_5O_7 \rightleftharpoons H_2C_6H_5O_7^- + H^+$ $\quad K_{a_1} = \dfrac{[H_2C_6H_5O_7^-][H^+]}{[H_3C_6H_5O_7]}$

$H_2C_6H_5O_7^- \rightleftharpoons HC_6H_5O_7^{2-} + H^+$ $\quad K_{a_2} = \dfrac{[HC_6H_5O_7^{2-}][H^+]}{[H_2C_6H_5O_7^-]}$

$HC_6H_5O_7^{2-} \rightleftharpoons C_6H_5O_7^{3-} + H^+$ $\quad K_{a_3} = \dfrac{[C_6H_5O_7^{3-}][H^+]}{[HC_6H_5O_7^{2-}]}$

16.71 (See Table 16.2) The second ionization will have no noticeable effect on the pH. We can ignore it. $\quad H_2C_6H_6O_6 \rightleftharpoons H^+ + HC_6H_6O_6^-$

M.W. $(H_2C_6H_6O_6) = 176$

$$M\,(H_2C_6H_6O_6) = 0.500\text{g} \times \frac{1\text{ mol}}{176\text{ g}} \times \frac{1}{0.250\text{ L}} = 1.14 \times 10^{-2}$$

$$K_a = 7.9 \times 10^{-5} = \frac{X^2}{1.14 \times 10^{-2} - X}$$

First approximation $X = [H^+] = 9.49 \times 10^{-4}$

Second approximation $X = [H^+] = 9.1 \times 10^{-4}$

pH = 3.04

16.73 $H_2C_6H_6O_6 \rightleftharpoons H^+ + HC_6H_6O_6^-$ $K_{a_1} = 7.9 \times 10^{-5}$

$HC_6H_6O_6^- \rightleftharpoons H^+ + C_6H_6O_6^{2-}$ $K_{a_2} = 1.6 \times 10^{-12}$

$$M(H_2C_6H_6O_6) = 0.500\text{ g} \times \frac{1\text{ mol}}{176\text{ g}} \times \frac{1}{0.200\text{ L}} = 1.42 \times 10^{-2}\text{M}$$

First Ionization

	Init. Conc.	Change	Equil. Conc.
H^+	0.1 (from pH)	+X	$0.1 + X \approx 0.1$
$HC_6H_6O_6^-$	0	+X	+X
$H_2C_6H_6O_6$	1.42×10^{-2}	-X	$(1.42 \times 10^{-2}) - X \approx 1.42 \times 10^{-2}$

The second ionization will have no noticeable effect on the concentrations.

$$K_{a_1} = \frac{(0.1)X}{1.42 \times 10^{-2}} = 7.9 \times 10^{-5}$$

$X = 1.1 \times 10^{-5}$

$$\text{fraction dissociated} = \frac{1.1 \times 10^{-5}}{1.42 \times 10^{-2}} = \mathbf{7.9 \times 10^{-4}}$$

16.75 First Ionization

	Init. Conc.	Change	Equil. Conc.
H^+	?	?	3.7×10^{-8} (from pH)
HCO_3^-	0	+X	X
H_2CO_3	2.6×10^{-2}	-X	$2.6 \times 10^{-2} - X$

$$K_{a_1} = 4.3 \times 10^{-7} = \frac{(3.7 \times 10^{-8})X}{(2.6 \times 10^{-2} - X)}$$

$\mathbf{X = [HCO_3^-] = 2.4 \times 10^{-2}\ mol/L}$

(continued)

16.75 (continued)

Second Ionization

	Init. Conc.	Change	Equil. Conc.
H^+	--	--	3.7×10^{-8}
CO_3^{2-}	0	+X	X
HCO_3^-	2.4×10^{-2}	-X	2.4×10^{-2} - X

$$K_{a_2} = 5.6 \times 10^{-11} = \frac{(3.7 \times 10^{-8})X}{(2.4 \times 10^{-2} - X)} = \frac{(3.7 \times 10^{-8})X}{2.4 \times 10^{-2}}$$

X = 3.6 x 10^{-5}

$[HCO_3^-]$ = 2.4 x 10^{-2} - 3.6 x 10^{-5} = **2.4 x 10^{-2} mol/L**

16.77 $SO_3^{2-} + H_2O \rightleftharpoons HSO_3^- + OH^-$

$$K_{b_1} = \frac{K_w}{K_{a_2}} = \frac{1.0 \times 10^{-14}}{1.0 \times 10^{-7}} = 1.0 \times 10^{-7}$$

$$K_{b_1} = \frac{X^2}{0.25 - X} \approx \frac{X^2}{0.25} = 1.0 \times 10^{-7}$$

X = $[OH^-]$= 1.6 x 10^{-4}

The second hydrolysis (hydrolysis of the HSO_3^-) will yield negligible amounts of OH^-. **pH = 10.20**

16.79 Yes - due to hydrolysis

16.81 (a) Since the titration involves a strong acid and a strong base, the **pH at the equivalence point will be 7.00.**

(b) $\dfrac{0.0200 \text{ mol H}^+ \text{ (from HNO}_3)}{1{,}000 \text{ mL}} \times 15.0 \text{ mL} \times \dfrac{1 \text{ mol OH}^- \text{ (required)}}{1 \text{ mol H}^+ \text{ (available)}}$

$\times \dfrac{1{,}000 \text{ mL KOH (soln.)}}{0.0100 \text{ mol OH}^-}$

= 30.0 mL KOH required to titrate to equivalence point

(c) $\text{mol H}^+ \text{ available} = \dfrac{0.0200 \text{ mol H}^+}{1{,}000 \text{ mL}} \times 15.0 \text{ mL} = 3.00 \times 10^{-4} \text{ mol H}^+$

$\text{mol OH}^- \text{ available} = \dfrac{0.0100 \text{ mol OH}^-}{1{,}000 \text{ mL}} \times 10.0 \text{ mL} = 1.00 \times 10^{-4} \text{ mol OH}^-$

Excess $H^+ = 3.00 \times 10^{-4} - 1.00 \times 10^{-4} = 2.00 \times 10^{-4}$ mol

$\dfrac{2.00 \times 10^{-4} \text{ mol}}{15.0 \text{ mL} + 10.0 \text{ mL}} \times \dfrac{1{,}000 \text{ mL}}{\text{L}} = 8.00 \times 10^{-3} \text{ M}$

pH = - log 8.00×10^{-3} = **2.10**

(d) $\text{mol H}^+ \text{ available} = \dfrac{0.0200 \text{ mol H}^+}{1{,}000 \text{ mL}} \times 15.0 \text{ mL} = 3.00 \times 10^{-4} \text{ mol H}^+$

$\text{mol OH}^- \text{ available} = \dfrac{0.0100 \text{ mol OH}^-}{1{,}000 \text{ mL}} \times 35.0 \text{ mL} = 3.50 \times 10^{-4} \text{ mol OH}^-$

Excess $OH^- = 3.50 \times 10^{-4} - 3.00 \times 10^{-4} = 5.0 \times 10^{-5}$ mol OH^- excess

$\dfrac{5.0 \times 10^{-5} \text{ mol OH}^-}{15.0 \text{ mL} + 35.0 \text{ mL}} \times \dfrac{1{,}000 \text{ mL}}{\text{L}} = 1.0 \times 10^{-3} \text{ M OH}^-$

pOH = 3.00 **pH = 11.00**

16.83 $HF + NaOH \rightleftharpoons NaF + H_2O$

(a) $\frac{0.200 \text{ mol HF}}{1{,}000 \text{ mL}} \times 50.0 \text{ mL} = 0.0100 \text{ mol HF}$

$\frac{0.100 \text{ mol NaOH}}{1{,}000 \text{ mL}} \times 5.0 \text{ mL} = 0.00050 \text{ mol NaOH}$

After the initial acid-base reaction, there will be 0.0100 - 0.0005 or 0.0095 mol HF and 0.0005 mol NaF or F^-. This is a buffer solution.

$HF \rightleftharpoons H^+ + F^-$

	Init. Conc.	Change	Equil. Conc.
H^+	~0	+X	X
F^-	0.00050 mol/0.055 L	+X	(0.00050/0.055) + X
HF	0.0095 mol/0.055 L	-X	(0.0095/0.055) - X

$$K_a = 6.5 \times 10^{-4} = \frac{(X)[(0.00050/0.055) + X]}{(0.0095/0.055) - X}$$

X =[H^+] = 6.8×10^{-3} M (from quadratic equation) **pH = 2.17**

(b) 0.0100 mol HF + 0.0050 mol NaOH yields 0.0050 mol excess HF and 0.0050 mol NaF. $HF \rightleftharpoons H^+ + F^-$

	Init. Conc.	Change	Equil. Conc.
H^+	~0	+X	X
F^-	0.0050 mol/0.100 L	+X	≈ 0.0050/0.100
HF	0.0050 mol/0.100 L	-X	≈ 0.0050/0.100

$$K_a = 6.5 \times 10^{-4} = \frac{(X)(0.050)}{(0.050)}$$

X =[H^+] = 6.5×10^{-4} **pH = 3.19** (Note: pH = pK_a at half neutralization)

(continued)

16.83 (continued)

(c) $0.0100 \text{ mol HF} \times \frac{1 \text{ mol NaOH}}{1 \text{ mol HF}} \times \frac{1 \text{ L NaOH}}{0.100 \text{ mol NaOH}} \times \frac{1{,}000 \text{ mL}}{\text{L}}$

$= 100 \text{ mL NaOH solution required.}$

$$0.0100 \text{ mol HF} \times \frac{1 \text{ mol F}^- \text{ (or NaF)}}{1 \text{ mol HF}} = 0.0100 \text{ mol F}^-$$

$$\frac{0.0100 \text{ mol F}^-}{50.0 \text{ mL} + 100 \text{ mL}} \times \frac{1{,}000 \text{ mL}}{\text{L}} = 0.0667 \text{ M F}^-$$

$$F^- + H_2O \rightleftharpoons HF + OH^-$$

	Init. Conc.	Change	Equil. Conc.
F^-	0.0667	-X	0.0667 - X $\approx$ 0.0667
OH^-	~0	+X	X
HF	0	+X	X

$$K_b = \frac{K_w}{K_a} = \frac{1.0 \times 10^{-14}}{6.5 \times 10^{-4}} \approx \frac{(X)(X)}{0.0667}$$

$X = [OH^-] = 1.0 \times 10^{-6}$

pOH = 6.00

pH = 8.00

[Note: If the concentration of OH^- produced by the dissociation of water is considered, the pH will be slightly less. That is, if the initial concentration of OH^- is taken as 1×10^{-7} rather than ≈ 0, a slightly different and more precise answer will be obtained.]

16.85 $?\text{mol H}^+ \text{ (initial)} = 50.0 \text{ mL acid} \times \frac{0.10 \text{ mol H}^+}{10^3 \text{ mL acid}} = \mathbf{5.0 \times 10^{-3} \text{ mol H}^+}$

If mol H^+ > mol OH^-, the **mol H^+ (XS) = mol H^+ - mol OH^-**

If mol OH^- > mol H^+, the **mol OH^- (XS) = mol OH^- - mol H^+**

$[H^+]$ = mol H^+/final total volume in liters

$[OH^-]$ = mol OH^-/final total volume in liters

point	mL base added:	mol base	M H^+	pH
1	0.00	0.00	0.10	1.00
2	10.00	0.0010	0.067	1.18
3	20.00	0.0020	0.043	1.37
4	30.00	0.0030	0.025	1.60
5	40.00	0.0040	0.011	1.95
6	45.00	0.0045	0.0053	2.28
7	49.00	0.0049	0.0010	3.00
8	50.00	0.0050	1.0×10^{-7}	7.00
9	51.00	0.0051	1.0×10^{-11}	11.00
10	55.00	0.0055	2.1×10^{-12}	11.68
11	60.00	0.0060	1.1×10^{-12}	11.96
12	70.00	0.0070	6.0×10^{-13}	12.22
13	80.00	0.0080	4.3×10^{-13}	12.36
14	90.00	0.0090	3.5×10^{-13}	12.46
15	100.00	0.0100	3.0×10^{-13}	12.52

These results will yield a plot very much like Figure 16.5 with an equivalence point at pH = 7.0 and 50.00 mL of NaOH added.

16.87 Acid and basic forms of an indicator differ in color ($HIn \rightleftharpoons H^+ + In^-$). Therefore, depending on the pH range of your indicator, it will change to HIn in acid solution and In^- in basic solution. This color change ideally should correspond to the equivalence point of the titration. If too much indicator is added, it may interfere with the endpoint because it will react with the base in the titration.

16.89 No. pH range for the color change is too low.

16.91 Using the Henderson-Hasselbalch equation:

$$pH = pK_a + \log \frac{[anion]}{[acid]}$$

$$7.0 = -\log(1 \times 10^{-5}) + \log \frac{[anion]}{[acid]}$$

$$2.0 = \log \frac{[anion]}{[acid]} \qquad \frac{[anion]}{100} = [acid] \qquad 100 = \frac{[anion]}{[acid]}$$

The solution will be green.

16.93 Is HCO_3^- an acid or a base? It is both an acid and a base, but as which is it the stronger?

$$K_a\,(HCO_3^-) = K_{a\,2}\,(H_2CO_3) = 5.6 \times 10^{-11}$$

$$K_b\,(HCO_3^-) = K_{hy}\,(HCO_3^-) = \frac{K_w}{K_{a\,1}\,(H_2CO_3)} = \frac{1.0 \times 10^{-14}}{4.3 \times 10^{-7}} = 2.3 \times 10^{-8}$$

Since it is a stronger base than it is an acid, let's ignore its acid properties and calculate its pH as a base.

$$K_b = 2.3 \times 10^{-8} = \frac{[OH^-][H_2CO_3]}{0.50} = \frac{X^2}{0.50}$$

$X = [OH^-] = 1.1 \times 10^{-4}$

pH = 10.04

(The above assumption is not very accurate. The actual pH can be calculated by using the formula $\left[H^+\right] = \sqrt{K_{a\,1} \times K_{a\,2}}$ Then, $[H^+]$ would be calculated to be 4.91×10^{-9} and pH would be 8.31. The derivation of the equation $\left[H^+\right] = \sqrt{K_{a\,1} \times K_{a\,2}}$ requires the simultaneous consideration of both equilibria and is derived in the solution to Exercise 16.95. pH = 10 will be used in this

(continued)

16.93 (continued)

solution. If you can derive the equation $\left[H^+\right] = \sqrt{K_{a\,1} \times K_{a\,2}}$, please use pH = 8.31.) Final pH = ?

	Init. Conc.	Effect of HCl	Change	Equil. Conc.
HCO_3^-	0.50	-0.05	+X	$0.45 + X \approx 0.45$ M
H^+	--	--	X	X
H_2CO_3	0	+0.05	-X	$0.05 - X \approx 0.05$ M

$$K_a = K_{a\,1} = \frac{(X)(0.45)}{(0.05)} = 4.3 \times 10^{-7}$$

$X = [H^+] = 5 \times 10^{-8}$ pH = 7.3

ΔpH = 7.3 - 10.0 = **-2.7** (or 7.3 - 8.31 = -1.0)

16.95 0.10 M NH_4^+ and 0.10 M NO_2^-

$$K_a\,(NH_4^+) = \frac{K_w}{K_b} = \frac{1.0 \times 10^{-14}}{1.8 \times 10^{-5}} = 5.6 \times 10^{-10}$$

$$K_b\,(NO_2^-) = \frac{K_w}{K_a} = \frac{1.0 \times 10^{-14}}{4.5 \times 10^{-4}} = 2.2 \times 10^{-11}$$

The values are too close to be able to neglect one or the other. You will need to consider both equations simultaneously. Consider what you already know.

(1) $NH_4^+ \rightleftharpoons NH_3 + H^+$ $K_a\,(NH_4^+) = \dfrac{K_w}{K_b}$

(continued)

16.95 (continued)

(2) $NO_2^- + H_2O \rightleftharpoons HNO_2 + OH^-$ $\qquad K_b(NO_2^-) = \frac{K_w}{K_a}$

(3) Initially, $[NH_4^+]^- = [NO_2^-]$ and this will probably not change significantly, since one is a very weak acid while the other is a very weak base.

(4) $[H^+][OH^-] = 1.0 \times 10^{-14}$ or $[OH^-] = 1.0 \times 10^{-14}/[H^+]$

Divide $K_a(NH_4^+)$ by $K_b(NO_2^-)$

$$\frac{K_a(NH_4^+)}{K_b(NO_2^-)} = \frac{[NH_3][H^+]/[NH_4^+]}{[HNO_2][OH^-]/[NO_2^-]}$$

Since the denominators are equal:

$$\frac{K_a(NH_4^+)}{K_b(NO_2^-)} = \frac{[NH_3][H^+]}{[HNO_2][OH^-]} = \frac{[NH_3][H^+]}{[HNO_2]\,K_w/[H^+]}$$

$$\frac{K_a(NH_4^+)}{K_b(NO_2^-)} = \frac{[NH_3][H^+]^2}{[HNO_2]\,K_w}$$

If $[NH_4^+] \approx [NO_2^-]$, then $[NH_3] \approx [HNO_2]$ and the equation becomes:

$$\frac{K_a(NH_4^+)}{K_b(NO_2^-)} = \frac{[H^+]^2}{K_w}$$

$$[H^+] = \sqrt{\frac{K_a(NH_4^+)\,K_w}{K_b(NO_2^-)}}$$

$$= \sqrt{K_a(NH_4^+)\,K_a(HNO_2)}$$

$$H^+ = \sqrt{5.6 \times 10^{-10} \times 4.5 \times 10^{-4}} = 5.0 \times 10^{-7}$$

pH = 6.30

16.97 $HIO_3 \rightleftharpoons H^+ + IO_3^-$

	HIO_3	H^+	IO_3^-
before	0.20	≈0	0
change	-X	+X	+X
equil	0.20 - X	X	X

$$K_a = 1.7 \times 10^{-1} = \frac{(X)(X)}{(0.20 - X)}$$

(Use the quadratic formula)

$X = 0.12$

$[H^+] = 0.12$

$[IO_3^-] = 0.12$

$[HIO_3] = 0.08$

pH = **0.92**

16.99 $PO_4^{3-} + H_2O \rightleftharpoons HPO_4^{2-} + OH^-$

	PO_4^{3-}	HPO_4^{2-}	OH^-
Before	1.0	0	≈0
change	-X	+X	+X
equil	1.0 - X	X	X

$$K_b = \frac{K_w}{K_{a_3}} = \frac{1.0 \times 10^{-14}}{2.2 \times 10^{-12}} = \frac{(X)(X)}{1.0 - X}$$

$$4.55 \times 10^{-3} = \frac{X^2}{1.0 - X}$$

by successive approximations

$4.55 \times 10^{-3} = X^2 \quad X = 6.75 \times 10^{-2}$

(continued)

16.99 (continued)

$$4.55 \times 10^{-3} = \frac{X^2}{1.0 - (6.75 \times 10^{-2})} \quad X = 6.51 \times 10^{-2}$$

$$4.55 \times 10^{-3} = \frac{X^2}{1.0 - (6.5 \times 10^{-2})} \quad X = 6.5 \times 10^{-2}$$

$X = 6.5 \times 10^{-2} = [OH^-]$ $\quad$ $pOH = 1.19$

pH = 12.81

$NaOH \rightarrow Na^+ + OH^-$

$[OH^-]$ = **6.5×10^{-2} M = conc. of NaOH needed to give the same pH.**

17 SOLUBILITY AND COMPLEX ION EQUILIBRIA

17.1 The concentration of the solid is left out of the solubility equilibrium expression of a salt because it is not a variable and can be (and is) included within the constant K_{sp}.

17.3 (a) $K_{sp} = [Pb^{2+}][F^-]^2$

(b) $K_{sp} = [Ag^+]^3[PO_4^{3-}]$

(c) $K_{sp} = [Fe^{2+}]^3[PO_4^{3-}]^2$

(d) $K_{sp} = [Li^+]^2[CO_3^{2-}]$

(e) $K_{sp} = [Ca^{2+}][IO_3^-]^2$

(f) $K_{sp} = [Ag^+]^2[Cr_2O_7^{2-}]$

17.5 $PbCO_3(s) \rightarrow Pb^{2+}(aq) + CO_3^{2-}(aq)$

$K_{sp} = [Pb^{2+}][CO_3^{2-}] = (1.8 \times 10^{-7})(1.8 \times 10^{-7}) = \mathbf{3.2 \times 10^{-14}}$

17.7 $CaCrO_4(s) \rightarrow Ca^{2+}(aq) + CrO_4^{2-}(aq)$

$K_{sp} = [Ca^{2+}][CrO_4^{2-}] = (1.0 \times 10^{-2})(1.0 \times 10^{-2}) = \mathbf{1.0 \times 10^{-4}}$

17.9 $\frac{0.0981\ g}{0.200\ L} \times \frac{1\ mol}{245.2\ g} = \frac{2.00 \times 10^{-3}\ mol}{L}$

$PbF_2(s) \rightarrow Pb^{2+}(aq) + 2F^-(aq)$

$K_{sp} = [Pb^{2+}][F^-]^2 = (2.00 \times 10^{-3})(2 \times 2.00 \times 10^{-3})^2 = \mathbf{3.20 \times 10^{-8}}$

17.11 $\frac{4.99\ g}{L} \times \frac{1\ mol}{312\ g} = \frac{1.6 \times 10^{-2}\ mol}{L}$

$Ag_2SO_4(s) \rightarrow 2Ag^+(aq) + SO_4^{2-}(aq)$

$K_{sp} = [Ag^+]^2[SO_4^{2-}] = (2 \times 1.6 \times 10^{-2})^2(1.6 \times 10^{-2}) = \mathbf{1.6 \times 10^{-5}}$

17.13 $\frac{0.47\ g\ MgC_2O_4}{0.500\ L} \times \frac{1\ mol}{112\ g} = 8.4 \times 10^{-3}\ M$

$MgC_2O_4(s) \rightarrow Mg^{2+}(aq) + C_2O_4^{2-}(aq)$

	Init. Conc.	Added	Equil. Conc.
Mg^{2+}	0	8.4×10^{-3}	8.4×10^{-3}
$C_2O_4^{2-}$	2.0×10^{-3}	8.4×10^{-3}	10.4×10^{-3}

$K_{sp} = (8.4 \times 10^{-3})(10.4 \times 10^{-3}) = \mathbf{8.7 \times 10^{-5}}$

17.15 $Mg(OH)_2(s) \rightarrow Mg^{2+}(aq) + 2OH^-(aq)$

	Init. Conc.	Change	Equil. Conc.
Mg^{2+}	0	+X	X
OH^-	1×10^{-7}	+2X	$1 \times 10^{-7} + 2X \approx 2X$

$K_{sp} = 7.1 \times 10^{-12} = (X)(2X)^2$ $X = 1.2 \times 10^{-4}$
$[OH^-] = 2.4 \times 10^{-4}$

pH = 10.38

17.17 Volume of plaster = $\pi r^2 h = 3.14 \times (0.5\text{ cm})^2(1.50\text{ cm}) = 1.18\text{ cm}^3$ (assuming that 0.5 cm has 3 significant figures)
Mass of plaster = $1.18\text{ cm}^3 \times 0.97\text{ g/mL} \times 1\text{ mL/cm}^3 = 1.1\text{ g}$

$CaSO_4(s) \rightarrow Ca^{2+}(aq) + SO_4^{2-}(aq)$ $K_{sp} = 2. \times 10^{-4} = (X)(X)$

X = molar solubility = 1.4×10^{-2} mol/L

$(1.4 \times 10^{-2}\text{ mol/L}) \times 136\text{ g/mol} = 1.9\text{ g/L}$

Liters required to dissolve 1.1 g: 1.1 g x 1L/1.9 g = 0.58 L or 580 mL

Time required: 0.58 L x 1d/2.0 L = **0.29 day or 7.0 hours**

17.19 In a solution, a precipitate will form only if the mixture is supersaturated (i.e., when the value of the ion product exceeds the value of the K_{sp}).

17.21 K_{sp} of $Fe(OH)_2 = 2 \times 10^{-15} = [Fe^{2+}][OH^-]^2 = (0.010)(X)^2$
$X^2 = 2 \times 10^{-15}/0.010 = 2 \times 10^{-13}$ $X = [OH^-] = 4 \times 10^{-7}$
pOH = 6.4 **pH = 7.6**

This is the minimum pH at which the value of the ion product will equal or exceed the solubility product constant.

17.23 (a) The only possible precipitate is $CaCO_3$ ($K_{sp} = 9 \times 10^{-9}$).

Ion prod. = $(0.025)(0.0050) = 1.2 \times 10^{-4}$

Since Ion Product > K_{sp}, **$CaCO_3$ will precipitate.**

(b) The only possible precipitate is $PbCl_2$ ($K_{sp} = 1.6 \times 10^{-5}$)
Ion Product = $(0.010)(0.060)^2 = 3.6 \times 10^{-5}$

Ion Product > K_{sp} **$PbCl_2$ will precipitate**

(c) The only possible precipitate is FeC_2O_4 ($K_{sp} = 2.1 \times 10^{-7}$)

Ion Product = $(1.5 \times 10^{-3})(2.2 \times 10^{-3}) = 3.3 \times 10^{-6}$

Ion Product > K_{sp} **FeC_2O_4 will precipitate.**

17.25 K_{sp} $(CaCO_3) = 9 \times 10^{-9} = [Ca^{2+}][CO_3^{2-}] = (X)(X + 0.50)$

$X = 2 \times 10^{-8}$ mol/L = molar solubility of $CaCO_3$ in 0.50 M Na_2CO_3

17.27 $K_{sp} = 1.6 \times 10^{-5} = [Pb^{2+}][Cl^-]^2 = (X)(2X + 0.060)^2$

$X = 4.4 \times 10^{-3}$ if $2X + 0.060$ is assumed to be ≈ 0.060.
Not a good assumption.

Assume $2X + 0.060 \approx 2(4.4 \times 10^{-3}) + 0.060$ or 0.069.

Then $1.6 \times 10^{-5} = (X)(0.069)^2$ or $X = 3.6 \times 10^{-3}$

3.6×10^{-3} mol/L is the calculated molar solubility of $PbCl_2$ in 0.020 M $AlCl_3$

17.29 $K_{sp} = 1.9 \times 10^{-12} = [Ag^+]^2[CrO_4^{2-}] = (2X)^2(X + 0.10) \approx (4X^2)(0.10)$

$X = 2.2 \times 10^{-6}$ mol Ag_2CrO_4/L in 0.10 M Na_2CrO_4

17.31 X = moles of NaF added per liter

K_{sp} (BaF_2) = [Ba^{2+}][F^-]^2 = $(6.8 \times 10^{-4})(2 \times 6.8 \times 10^{-4} + X)^2 = 1.7 \times 10^{-6}$

(from the solution of the quadratic equation) X = 4.9×10^{-2} mol NaF/L

$$\frac{4.9 \times 10^{-2}\ \text{mol}}{\text{L}} \times \frac{42\ \text{g}}{\text{mol}} = \mathbf{2.1\ g\ NaF}$$

17.33 $Ca(OH)_2 \rightleftharpoons Ca^{2+} + 2OH^-$

(a) $K_{sp} = 6.5 \times 10^{-6} = (0.10 + X)(2X)^2$
X = 4.0×10^{-3} = molar solubility of $Ca(OH)_2$
= **4.0 x 10^{-3} mole/L**
(b) $K_{sp} = 6.5 \times 10^{-6} = (X)(2X + 0.10)^2$
X = 6.5×10^{-4} = molar solubility of $Ca(OH)_2$
= **6.5 x 10^{-4} mole/L**

17.35 M NaOH = (2.20 g/40.0 g/mol)/0.250 L = 0.220 M

	Init. Conc.	Change	Equil. Conc.
Na^+	0.220 M	-	- - -
OH^-	0.220 M	-2X	0.220 - 2X
Fe^{2+}	0.10 M	-X	0.10 - X
Cl^-	0.20 M	-	- - -

Assume that X ≈ 0.100; then [OH^-] ≈ 0.020 and [Fe^{2+}] = Y

$K_{sp} = 2 \times 10^{-15} = [Fe^{2+}][OH^-]^2 = (Y)(0.020)^2$

Y = 5.0×10^{-12} **[Fe^{2+}] = 5 x 10^{-12} M**

The amount $Fe(OH)_2$ formed will be: [0.10 mol/L - (5×10^{-12} mol/L)] x 0.250 L x 89.9 g/mol = **2.2 g $Fe(OH)_2$ precipitated**

17.37 K_{sp} $Mn(OH)_2 = 1.2 \times 10^{-11}$ K_{sp} $Fe(OH)_2 = 2 \times 10^{-15}$

	Init. Conc.	Added	Change	Equil. Conc.	Assume	Equil. Conc.
Fe^{2+}	0.100		-Y	0.100 - Y	$Y \approx 0.100$	Fe^{2+}
Mn^{2+}	0	X		X	$X \approx 0.100$	0.100
OH^-	1×10^{-7}	2X	-2Y	2X - 2Y		OH^-

K_{sp} $(Mn(OH)_2) = 1.2 \times 10^{-11} = (0.10)[OH^-]^2$, $[OH^-] = 1.1 \times 10^{-5}$

K_{sp} $(Fe(OH)_2) = 2 \times 10^{-15} = [Fe^{2+}](1.1 \times 10^{-5})^2$

$[Fe^{2+}] \approx 2 \times 10^{-5}$ M

pH = 14 + log 1.1 x 10^{-5} = **9.04**

$[Mn^{2+}] \approx 0.100$ M

17.39 $K_{a_1} \times K_{a_2} = 4.0 \times 10^{-6} = \dfrac{[H^+]^2[C_2O_4^{2-}]}{[H_2C_2O_4]}$

For maximum separation: ion product MgC_2O_4 equals K_{sp} (MgC_2O_4)

$8.6 \times 10^{-5} = [Mg^{2+}][C_2O_4^{2-}] = (0.10\ M)[C_2O_4^{2-}]$

$[C_2O_4^{2-}] = 8.6 \times 10^{-4}$ M

$$4.0 \times 10^{-6} = \frac{[H^+]^2 [C_2 O_4^{2-}]}{[H_2 C_2 O_4]} = \frac{[H^+]^2(8.6 \times 10^{-4})}{(0.10)}$$

$[H^+] = 2.2 \times 10^{-2}$, pH = 1.66

pH of less than 1.66

17.41 $H_2CO_3 \rightleftharpoons 2H^+ + CO_3^{2-}$

$K_a = K_{a_1} \times K_{a_2}$ = (from Table 16.2) $4.3 \times 10^{-7} \times 5.6 \times 10^{-11} = 2.4 \times 10^{-17}$

Maximum CO_3^{2-} before precipitation of $CaCO_3$
$K_{sp} = 9 \times 10^{-9} = (0.050)[CO_3^{2-}]$
$[CO_3^{2-}] = 2 \times 10^{-7}$

Maximum CO_3^{2-} before precipitation of $PbCO_3$
$K_{sp} = 7.4 \times 10^{-14} = (0.050)[CO_3^{2-}]$
$[CO_3^{2-}] = 1.5 \times 10^{-12}$

Maximum separation will be when $[CO_3^{2-}]$ = slightly less than 2×10^{-7}

$H_2CO_3 \rightleftharpoons 2H^+ + CO_3^{2-}$

$$K_a = 2.4 \times 10^{-17} = \frac{[H^+]^2(2 \times 10^{-7})}{0.050}$$

$[H^+] = 2.4 \times 10^{-6}$

Less than pH = 5.62

17.43 (a) $Fe(CN)_6^{4-} \rightleftharpoons Fe^{2+} + 6CN^-$

$$K_{inst} = \frac{[Fe^{2+}][CN^-]^6}{[Fe(CN)_6^{4-}]}$$

(b) $$K_{inst} = \frac{[Cu^{2+}][Cl^-]^4}{[CuCl_4^{2-}]}$$

(c) $$K_{inst} = \frac{[Ni^{2+}][NH_3]^6}{[Ni(NH_3)_6^{2+}]}$$

17.45 $AgI_2^-(aq) \rightleftharpoons Ag^+(aq) + 2I^-(aq)$
A more dilute solution would favor the right side of this equilibrium. The presence of the ions, Ag^+ and I^-, would in turn favor the formation of solid AgI via
$Ag^+(aq) + I^-(aq) \rightarrow AgI(s)$.

17.47 K_{sp} $(Zn(OH)_2) = 1.2 \times 10^{-23}$ $Zn(NH_3)_4^{2+} \rightleftharpoons Zn^{2+} + 4NH_3$ $K_{inst} = ?$

	Init. Conc.	Change	Equil. Conc.	Assume	Equil. Conc.
Zn^{2+}	5.7×10^{-3}	-X	5.7×10^{-3} - X	$X \approx 5.7 \times 10^{-3}$	$[Zn^{2+}]$
OH^-	$2 \times 5.7 \times 10^{-3}$	0	0.0114		0.0114
NH_3	1.0	-4X	1.0 - 4X	≈ 1.0	1.0
$Zn(NH_3)_4^{2+}$	0	+X	X		5.7×10^{-3}

Using K_{sp}: $1.2 \times 10^{-23} = [Zn^{2+}](0.0114)^2$

$[Zn^{2+}] = 9.2 \times 10^{-20}$

$$K_{inst} = \frac{(9.2 \times 10^{-20})(1.0)^4}{5.7 \times 10^{-3}} = \mathbf{1.6 \times 10^{-17}}$$

[If one uses K_{sp} $(Zn(OH)_2) = 4.5 \times 10^{-17}$, a value of $K_{inst} = 6.1 \times 10^{-11}$ will be obtained.]

17.49 $AgI(s) + I^-(aq) \rightleftharpoons AgI_2^-(aq)$

$K_c = K_{sp} \times K_{form} = 8.5 \times 10^{-17} \times 1.0 \times 10^{11} = 8.5 \times 10^{-6}$

	Init. Conc.	Change	Equil.
I^-	1.0 M	-X	1.0 -X
AgI_2^-	0	+X	X

$$K_c = 8.5 \times 10^{-6} = \frac{[AgI_2^-]}{[I^-]} = \frac{X}{1.0 - X}$$

$X = 8.5 \times 10^{-6}$

8.5×10^{-6} mol/L of AgI will dissolve in 1.0 M NaI.

17.51 $AgC_2H_3O_2(s) \rightleftharpoons Ag^+ + C_2H_3O_2^-$ $K_{sp} = 2.3 \times 10^{-3}$

$HC_2H_3O_2 \rightleftharpoons H^+ + C_2H_3O_2^-$ $K_a = 1.8 \times 10^{-5}$

$$1.8 \times 10^{-5} = \frac{[H^+][[C_2H_3O_2^-]}{[HC_2H_3O_2]} = \frac{[C_2H_3O_2^-]^2}{1.0}$$

$[C_2H_3O_2^-] = 4.2 \times 10^{-3}$ Ion Product for the possible precipitate $AgC_2H_3O_2$ is: $1.0 \times 4.2 \times 10^{-3} = 4.2 \times 10^{-3}$. The Ion Product exceeds the value of K_{sp}; therefore, **a precipitate will form.**

17.53 $K_{sp} = [Ag^+][C_2H_3O_2^-] = 2.3 \times 10^{-3} = (0.200)[C_2H_3O_2^-]$

$[C_2H_3O_2^-] = 1.2 \times 10^{-2}$ $HC_2H_3O_2 \rightleftharpoons H^+ + C_2H_3O_2^-$

$[H^+] = 0.10 - X$ $[HC_2H_3O_2] = X$

$$1.8 \times 10^{-5} = \frac{(0.10 - X)(1.2 \times 10^{-2})}{X}$$ $X = 9.98 \times 10^{-2}$ or 0.10 M

Amount of $NaC_2H_3O_2$ required would be

> [0.10 mol/L + 0.012 mol/L] x 0.200 L x 82 g/mol **> 1.8 g of $NaC_2H_3O_2$**

17.55 $NH_3 + H_2O \rightleftharpoons NH_4^+ + OH^-$

$$K_b = 1.8 \times 10^{-5} = \frac{[OH^-]^2}{0.10}$$

$[OH^-]$ from $NH_3 = 1.3 \times 10^{-3}$ M

	$Mg(OH)_2(s)$ →	$Mg^{2+}(aq)$ +	$2OH^-(aq)$
Initial		0	1.3×10^{-3}
Change		+X	+2X
Equil.		X	$2X + 1.3 \times 10^{-3}$

$K_{sp} = [Mg^{2+}][OH^-]^2 = 7.1 \times 10^{-12} = (X)(2X + 1.3 \times 10^{-3})^2$

$\approx (X)(1.3 \times 10^{-3})^2$

X = molar solubility of $Mg(OH)_2$ = 4.2×10^{-6} mol/L

17.57 $CuSO_4 + 4NH_3 \rightleftharpoons Cu(NH_3)_4^{2+} + SO_4^{2-}$

$CuSO_4 + 2NH_3 + 2H_2O \rightleftharpoons Cu(OH)_2 + 2NH_4^+ + SO_4^{2-}$

Assume that the maximum amount of $Cu(NH_3)_4^{2+}$ will be formed.

	Cu^{2+}	+ $4NH_3$	$\rightleftharpoons$ $Cu(NH_3)_4^{2+}$
Initial	0.050	0.50	0
Change	-X	-4X	+X
Equil.	0.050 - X	0.50 - 4X	X (Assume X = 0.050)
Assume	Y	0.50 - 0.20 + 4Y	0.050 - Y ≈ 0.050

$$K_{form} = 4.8 \times 10^{12} = \frac{0.050}{Y(.30)^4}$$

$Y = [Cu^{2+}] = 1.3 \times 10^{-12}$

For formation of OH^-, look at K_b of NH_3

$$NH_3 + H_2O \rightleftharpoons NH_4^+ + OH^-$$

	NH_3	NH_4^+	OH^-
Initial (after complex)	.30	0	≈0
Change	-X	+X	+X
Equil.	.30 - X	X	X

$$K_b = 1.8 \times 10^{-5} = \frac{X^2}{.30 - X} = \frac{X^2}{.30} \qquad X = [OH^-] = 2.3 \times 10^{-3}$$

For $Cu(OH)_2$

	$Cu(OH)_{2(s)} \rightleftharpoons$ Cu^{2+}	+ $2OH^-$
Initial	1.3×10^{-12}	2.3×10^{-3}
Change	$-1.3 \times 10^{-12} + X$	$-2(1.3 \times 10^{-12}) + 2X$
Equil.	X	$\approx 2.3 \times 10^{-3}$

$4.8 \times 10^{-20} = (X)(2.3 \times 10^{-3})^2$

$X = [Cu^{2+}] = 9.1 \times 10^{-15}$ M

18 ELECTROCHEMISTRY

18.1 Electrochemistry is the study of the relationships that exist between chemical reactions and the flow of electricity. An electrochemical reaction is either the nonspontaneous redox reaction caused by the passage of electricity through a system or the spontaneous redox reaction that is able to supply electricity.

18.3 In order for electrolytic conduction to occur in an aqueous solution of an electrolyte there must be oxidation at one electrode and reduction at the other electrode.

18.5 Prepare a sketch similar to that shown in Figure 18.3, but make the following changes:
(a) Replace the Na with Mg.
(b) Replace the Na^+ with Mg^{2+}.
(c) The half-reaction at the cathode will involve two electrons.
(d) Show electrons flowing from the anode to the D.C. source and from the D.C. source to the cathode.

18.7 The products would be O_2 at the anode due to the oxidation of water and H_2 at the cathode due to the reduction of the hydrogen ion.

18.9 anode: $2I^- \rightarrow I_2 + 2e^-$
cathode: $Ni^{2+} + 2e^- \rightarrow Ni$

cell: $Ni^{2+} + 2I^- \rightarrow Ni + I_2$
or $NiI_2(aq) \rightarrow Ni(s) + I_2(aq)$

18.11 (a) cathode: $2H_2O + 2e^- \rightarrow H_2 + 2OH^-$
anode: $2Cl^- \rightarrow Cl_2 + 2e^-$

net: $2H_2O + 2Cl^- \rightarrow H_2 + Cl_2 + 2OH^-$

(b) In a stirred solution, Cl_2 reacts with OH^-

$$Cl_2 + 2OH^- \rightarrow Cl^- + OCl^- + H_2O$$

Net reaction is: $Cl^- + H_2O \rightarrow OCl^- + H_2$

18.13 The advantage of the mercury electrolysis cell for production of NaOH and Cl_2 is that the NaOH obtained is not contaminated by Cl^-. The disadvantage is the possibility for mercury pollution.

18.15 Cryolite reduces the melting point of Al_2O_3 from 2000°C to about 1000°C which is low enough to make the electrolysis of molten Al_2O_3 feasible.

18.17 $Mg^{2+}(aq)$ (from sea water) + $2OH^-(aq) \rightarrow Mg(OH)_2(s)$
$Mg(OH)_2(s) + 2HCl \rightarrow MgCl_2(s) + 2H_2O$
$MgCl_2(\ell) \xrightarrow[\text{energy}]{\text{electrical}} Mg(\ell) + Cl_2(g)$

18.19 Electroplating is the process by which a metal is caused to be deposited ("plated out") on an electrode in an electrolysis cell. To have nickel plated onto an object in a $NiSO_4$ solution, the object must be part of the cathode because the Ni^{2+} must be reduced to Ni metal.

18.21 A faraday is the amount of electricity equal to the charge of 1 mole of electrons (96,500 coulombs).

18.23 $1\ C \times \frac{1\ F}{96{,}500\ C} \times \frac{6.022 \times 10^{23}\ e^-}{1\ F} = \mathbf{6.24 \times 10^{18}\ e^-}$

18.25 (a) $8950\ C \times \frac{1\ mol\ e^-}{96{,}500\ C} \Leftrightarrow \mathbf{0.0927\ mol\ e^-}$

(b) $1.5\ A \times 30\ s \times \frac{1\ C}{A \times s} \times \frac{1\ mol\ e^-}{96{,}500\ C} \Leftrightarrow \mathbf{4.7 \times 10^{-4}\ mol\ e^-}$

(c) $14.7\ A \times 10\ min \times \frac{60\ s}{min} \times \frac{1\ C}{A \times s} \times \frac{1\ mol\ e^-}{96{,}500\ C} \Leftrightarrow \mathbf{9.1 \times 10^{-2}\ mol\ e^-}$

18.27 (a) $\frac{84{,}200\ C}{6.30\ A} \times \frac{A \times s}{1\ C} \times \frac{1\ min}{60.0\ s} \Leftrightarrow \mathbf{223\ min}$

(b) $\frac{1.25\ mol\ e^-}{8.40\ A} \times \frac{96{,}500\ C}{1\ mol\ e^-} \times \frac{A \times s}{C} \times \frac{1\ min}{60.0\ s} \Leftrightarrow \mathbf{239\ min}$

(c) $\frac{0.500\ mol\ Al}{18.3\ A} \times \frac{3 \times 96{,}500\ C}{1\ mol\ Al} \times \frac{A \times s}{1\ C} \times \frac{1\ min}{60.0\ s} \Leftrightarrow \mathbf{132\ min}$

18.29 $25\ A \times 8.0\ hr \times \frac{3600\ s}{1\ hr} \times \frac{1\ C}{A \times s} \times \frac{F}{96{,}500\ C} \times \frac{1\ mol\ Na}{F} \times \frac{23.0\ g\ Na}{mol\ Na} \Leftrightarrow \mathbf{170\ g\ Na}$

Similar calculation yields: **260 g Cl_2** (only 2 significant figures)

18.31 $115\ A \times 8.00\ hr \times \frac{1\ C}{A \times s} \times \frac{3600\ s}{1\ hr} \times \frac{F}{96{,}500\ C} \times \frac{1\ mol\ Cu}{2F} \times \frac{63.55\ g\ Cu}{mol}$

$\Leftrightarrow$ **1,090 g Cu** (3 significant figures)

18.33 $\frac{21.4 \text{ g Ag}}{10.0 \text{ A}} \times \frac{\text{A x s}}{1 \text{ C}} \times \frac{96{,}500 \text{ C}}{F} \times \frac{1\, F}{\text{mol Ag}} \times \frac{1 \text{ mol Ag}}{107.9 \text{ g Ag}} \Leftrightarrow$ **1910 s**

(3 significant figures)

18.35 $\frac{5.00 \text{ g Cu}}{5.00 \text{ A}} \times \frac{\text{A x s}}{1 \text{ C}} \times \frac{2 \text{ mol } e^-}{63.55 \text{ g Cu}} \times \frac{96{,}500 \text{ C}}{\text{mol } e^-} \times \frac{1 \text{ min}}{60 \text{ s}} \Leftrightarrow$ **50.6 min**

18.37 $\frac{1.33 \text{ g } Cl_2}{45.0 \text{ min}} \times \frac{2 \text{ mol } e^-}{70.9 \text{ g } Cl_2} \times \frac{96{,}500 \text{ C}}{\text{mol } e^-} \times \frac{1 \text{ min}}{60 \text{ s}} \times \frac{\text{A x s}}{1 \text{ C}} \Leftrightarrow$ **1.34 A**

18.39 $0.500 \text{ L} \times \frac{0.270 \text{ mol } Cr_2(SO_4)_3}{1 \text{ L}} \times \frac{6 \text{ mol } e^-}{1 \text{ mol } Cr_2(SO_4)_3} \times \frac{96{,}500 \text{ C}}{\text{mol } e^-}$

$\times \frac{\text{A x s}}{\text{C}} \times \frac{1}{3.00 \text{ A}} \times \frac{1 \text{ min}}{60 \text{ s}} \Leftrightarrow$ **434 min**

18.41 $0.125 \text{ mol Cu} \times \frac{2 \text{ mol } e^-}{1 \text{ mol Cu}} \times \frac{1 \text{ mol Cr}}{3 \text{ mol } e^-} \Leftrightarrow$ **0.0833 mol Cr**

18.43 The electron flow takes place on the surface of the zinc. Electrons are removed from the zinc and are picked up by the Cu^{2+} ions that are in the vicinity. The energy change appears as heat.

18.45 In galvanic cells the anode is negative and the cathode is positive. In electrolytic cells the anode is positive and the cathode is negative.

18.47 In a galvanic cell, the negative electrode is the anode; the positive electrode is the cathode. These can be determined with a voltmeter. Another method would be to do a chemical analysis of the products formed at each electrode. Oxidation occurs at the anode, reduction at the cathode.

18.49 (a) $Zn(s) + Pb^{2+} (1M) \rightarrow Pb(s) + Zn^{2+} (1\ M)$
Zn is the anode; Pb is the cathode.
(b) $Al(s) + 3Ag^{+} (1\ M) \rightarrow Al^{3+} (1\ M) + 3Ag(s)$
Al is the anode; Ag is the cathode.
(c) $3Mn(s) + 2Fe^{3+} (1\ M) \rightarrow 3Mn^{2+} (1\ M) + 2Fe(s)$
Mn is the anode; Fe is the cathode.

18.51 (a) (anode) $Zn(s) \rightarrow Zn^{2+} + 2e^-$ (cathode) $Ga^{3+} + 3e^- \rightarrow Ga(s)$
(b) $3Zn(s) + 2Ga^{3+} \rightarrow 3Zn^{2+} + 2Ga(s)$
(c) $E_{cell} = E^\circ_{Ga^{3+}|Ga} - E^\circ_{Zn^{2+}|Zn} = E^\circ_{Ga^{3+}|Ga} - (-0.76\ V) = 0.23\ V$
$E^\circ_{Ga^{3+}|Ga} = \mathbf{-0.53\ V}$
(d) $E^\circ_{cell} = 0.34\ V - (-0.53\ V) = \mathbf{0.87\ V}$
(One must assume $Cu^{2+} + 2e^- \rightarrow Cu$)

18.53 (a) ClO_3^- (b) $Cr_2O_7^{2-}$ (c) MnO_4^- (d) PbO_2

18.55 (a) Na (b) Cl_2 (c) Cu (d) Sn (e) H_2

18.57 (a) $E_{cell} = -0.13\ V - (-0.76\ V) = \mathbf{0.63\ V}$
(b) $E_{cell} = 0.80\ V - (-1.67\ V) = \mathbf{2.47\ V}$
(c) $E_{cell} = -0.04\ V - (-1.03\ V) = \mathbf{0.99\ V}$

18.59 Spontaneous reactions are (a), (d), (e). The other reactions are spontaneous in the reverse direction.

18.61 (a) $Pb(s) + SO_4^{2-} + Hg_2Cl_2(s) \rightarrow 2Hg(\ell) + 2Cl^- + PbSO_4(s)$
$E^\circ = 0.27 - (-0.36) = \mathbf{0.63\ V}$
(b) $2Ag(s) + 2Cl^- + Cu^{2+} \rightarrow 2AgCl(s) + Cu(s)$ $E^\circ = \mathbf{0.12\ V}$
(c) $Mn(s) + Cl_2(g) \rightarrow Mn^{2+} + 2Cl^-$ $E^\circ = \mathbf{2.39\ V}$
(d) $2Al(s) + 3Br_2(\ell) \rightarrow 2Al^{3+} + 6Br^-$ $E^\circ = \mathbf{2.76\ V}$

18.63 (a) $E° = 0.77\ V - (-0.14\ V) = \mathbf{0.91\ V}$
(b) $E° = 0.00\ V - (0.34\ V) = \mathbf{-0.34\ V}$
(c) $E° = -2.38\ V - (-1.67\ V) = \mathbf{-0.71\ V}$
(d) $E° = -0.76\ V - (-1.03\ V) = \mathbf{0.27\ V}$
(e) $E° = 1.69\ V - (0.27\ V) = \mathbf{1.42\ V}$

18.65 (a) $E° = 0.77\ V - (-0.14\ V) = 0.91\ V$
$\log K_c = nE°/0.0592$ (at 25° C) $n = 2$
$\log K_c = 2 \times 0.91/0.0592 = 30.74$
$K_c = \mathbf{5 \times 10^{30}}$
(b) $K_c = \mathbf{3 \times 10^{-12}}$ (c) $K_c = \mathbf{1 \times 10^{-72}}$
(d) $K_c = \mathbf{1 \times 10^{9}}$ (e) $K_c = \mathbf{1 \times 10^{48}}$

18.67 $\Delta G° = -nF\,E°$

(a) $$\Delta G° = -2\ \text{mol} \times \frac{96{,}500\ \text{C}}{1\ \text{mol}\ e^-} \times 0.91\ \text{V} \times \frac{10^{-3}\ \text{kJ}}{1\ \text{V} \times \text{C}} = \mathbf{-180\ kJ}$$ (2 sign. fig.)

(b) $\Delta G° = -2 \times 96{,}500 \times (-0.34) \times 10^{-3} = \mathbf{66\ kJ}$
(c) $\Delta G° = -6 \times 96{,}500 \times (-0.71) \times 10^{-3} = \mathbf{410\ kJ}$
(d) $\Delta G° = -2 \times 96{,}500 \times 0.27 \times 10^{-3} = \mathbf{-52\ kJ}$
(e) $\Delta G° = -2 \times 96{,}500 \times 1.42 \times 10^{-3} = \mathbf{-274\ kJ}$

18.69 Nernst Equation: $\mathbf{E = E° - \frac{0.0592}{n} \times \log\ (mass\ action\ expression)}$
(a) $E° = 0.34\ V - (-0.76\ V) = \mathbf{1.10\ V}$
$E = 1.10\ V - (0.0592/2) \log \{[Zn^{2+}]/[Cu^{2+}]\} = \mathbf{1.07\ V}$
(b) $E° = -0.14\ V - (-0.25\ V) = \mathbf{0.11 V}$
$E = 0.11\ V - (0.0592/2) \log \{[Ni^{2+}]/[Sn^{2+}]\} = \mathbf{0.16\ V}$
(c) $E° = 2.87\ V - (-3.05\ V) = \mathbf{5.92\ V}$
$E = 5.92\ V - (0.0592/2) \log \left([Li^+]^2 [F^-]^2/p_{F_2}\right) = \mathbf{5.94\ V}$
(d) $E° = 0.00\ V - (-0.76\ V) = \mathbf{0.76\ V}$
$E = 0.76\ V - (0.0592/2) \log \left([Zn^{2+}]\ p_{H_2}/[H^+]^2\right) = \mathbf{0.64\ V}$
(e) $E° = 0.00\ V - (-0.44\ V) = \mathbf{0.44\ V}$
$E = 0.44\ V - (0.0592/2) \log \left(p_{H_2}[Fe^{2+}]/[H^+]^2\right) = \mathbf{0.46\ V}$

18.71 (a) $E =[-0.13\ V - (-0.14\ V)] - (0.0592/2) \log [(1.50)/(0.050)] = \mathbf{-0.03\ V}$

(b) $E =[-0.74\ V - (-0.76\ V)] - (0.0592/6) \log [(0.020)^3/(0.010)^2] = \mathbf{0.03\ V}$

(c) $E = (1.69\ V - 0.34\ V) - (0.0592/2) \log [(0.0010)/(0.010)\ (0.10)^4] = \mathbf{1.26\ V}$

18.73 $E° = 1.33\ V - (1.49\ V) = -0.16\ V$

$\Delta G = \Delta G° + 2.303\ RT\ \ \log ([MnO_4^-]^6\ [Cr^{3+}]^{10}/\{[Mn^{2+}]^6\ [Cr_2O_7^{2-}]^5\ [H^+]^{22}\})$

$$\Delta G° = -nF\ E° = -30 \text{ mol } e^- \times \frac{96{,}500\ C}{1 \text{ mol } e^-} \times (-0.16\ V) \times \frac{10^{-3}\ kJ}{1\ V \times C} = 460\ kJ$$

$$\Delta G = 460\ kJ + 2.303 \times 8.314\ J \times K^{-1} \times 298\ K \times \frac{1\ kJ}{10^3\ J}$$

$$\times \log \frac{(0.0010)^6 (0.0010)^{10}}{(0.10)^6\ (0.010)^5\ (1.0 \times 10^{-6})^{22}} = 460\ kJ + 5.71\ kJ \times \log 10^{100} = \mathbf{1.0 \times 10^3\ kJ}$$

Since the reaction as written involves free energy increase, the spontaneous reaction would be the reverse reaction.

18.75 Anode: Cl^- (?M) + Ag $\rightarrow$ AgCl + 1 e^-

Cathode: AgCl + 1 e^- $\rightarrow$ Ag + Cl^- (1 M)

Cl^- (? M) $\rightarrow$ Cl^- (1 M) A Concentration Cell

$E = 0.0435\ V$ $E° = 0$ $E = E° - (0.0592/1) \log \{(1)/[Cl^-]_A\}$

$0.0435\ V/(0.0592\ V) = - \log (1/[Cl^-]_A)$

$[Cl^-]_A = \mathbf{5.43\ M}$

18.77 $$25.0 \text{ g Pb} \times \frac{2 \times 96{,}500\ C}{207.2 \text{ g Pb}} \times \frac{1.5\ V}{25\ W} \times \frac{1\ W}{V \times C \times s^{-1}} \times \frac{1\ hr}{3600\ s} = \mathbf{0.39\ hr}$$

18.79 $5.00 \text{ min} \times 110 \text{ V} \times 1.00 \text{ A} \times \frac{1 \text{ W}}{\text{V} \times \text{A}} \times \frac{1 \text{ kJ}}{10^3 \text{ W} \times \text{s}} \times \frac{60 \text{ s}}{1 \text{ min}} = \mathbf{33.0\ kJ}$

18.81 $E = E° - \frac{0.0592}{2} \log \frac{[Zn^{2+}]}{[Pb^{2+}]}$

$0.4438 \text{ V} = 0.6365 \text{ V} - \frac{0.0592}{2} \log \frac{.500}{[Pb^{2+}]}$

$[Pb^{2+}] = \mathbf{1.6 \times 10^{-7}\ M}$

18.83 (anode) $Zn(s) \rightarrow Zn^{2+} + 2e^-$

(cathode) $2MnO_2(s) + 2NH_4^+ + 2e^- \rightarrow Mn_2O_3(s) + 2NH_3 + H_2O$

18.85 (anode) $Zn(s) + 2OH^- \rightarrow Zn(OH)_2(s) + 2e^-$

(cathode) $2MnO_2(s) + 2H_2O + 2e^- \rightarrow 2MnO(OH)(s) + 2OH^-$

---

net: $Zn(s) + 2MnO_2(s) + 2H_2O \rightarrow Zn(OH)_2(s) + 2MnO(OH)(s)$

Electrolyte is aqueous KOH.

18.87 (anode) $Zn(s) + 2OH^-(aq) \rightarrow Zn(OH)_2(s) + 2e^-$

(cathode) $Ag_2O(s) + H_2O + 2e^- \rightarrow 2Ag(s) + 2OH^-(aq)$

---

net: $Zn(s) + Ag_2O(s) + H_2O \rightarrow Zn(OH)_2(s) + 2Ag(s)$

18.89 (anode) $Cd(s) + 2OH^-(aq) \rightarrow Cd(OH)_2(s) + 2e^-$

(cathode) $NiO_2(s) + 2H_2O + 2e^- \rightarrow Ni(OH)_2(s) + 2OH^-(aq)$

---

net: $Cd(s) + NiO_2(s) + 2H_2O \rightarrow Cd(OH)_2(s) + Ni(OH)_2(s)$

18.91 Fuel cells operate under more nearly reversible conditions. Therefore, their thermodynamic efficiency is higher; i.e., more of the available energy can be used to do work.

18.93

	Init	STP
P	767 -27 torr	760 torr
V	288 mL	? mL
T	300 K	273 K

$$288 \text{ mL} \times \frac{740 \text{ torr} \times 273 \text{ K}}{760 \text{ torr} \times 300 \text{ K}} = \mathbf{255\ mL\ (STP)}$$

$$\frac{1.22 \text{ C} \times \text{s}^{-1} \times 30.0 \text{ min}}{255 \text{ mL } H_2 \text{ (STP)}} \times \frac{60 \text{ s}}{1 \text{ min}} \times \frac{11{,}200 \text{ mL } H_2}{6{,}022 \times 10^{23} \text{ e}^-} \Leftrightarrow 1.60 \times 10^{-19} \text{ C/e}^-$$

18.95 $$\frac{[H^+][C_2H_3O_2^-]}{[HC_2H_3O_2]} = 1.8 \times 10^{-5} = \frac{[H^+]^2}{0.10}$$

$[H^+]^2 = 1.8 \times 10^{-6}$

$2H^+ + Fe \rightarrow H_2 + Fe^{2+}$ $E = E° - (0.0592/n) \log \{[Fe^{2+}]/[H^+]^2\}$

$E = [(0.00 \text{ V} - (-0.44 \text{ V})] - (0.0296) \log \{(0.10)/(1.8 \times 10^{-6})\} = \mathbf{0.30\ V}$

18.97 $Zn(s) + 2Ag^{+}\ (0.0500\ M) \rightarrow Zn^{2+}\ (0.100\ M) + 2Ag(s)$ 45°C
(from Appendix C)

$\Delta H^{\circ} = -153.9\ kJ + 0\ kJ - (0\ kJ + 105.58\ kJ) = -259.48 \approx -259.5\ kJ$

$\Delta S^{\circ} = -112.1\ J/K + 42.55\ J/K - (41.6\ J/K + 72.68\ J/K) = -183.83 \approx -183.8\ J/K$

$$\Delta G^{\circ}_{318} = -259.5\ kJ - (318\ K)(-183.8\ J/K)\left(\frac{kJ}{10^{3}\ J}\right) = -259.5\ kJ + 58.4\ kJ = -201.1\ kJ$$

$$E^{\circ}_{318} = \frac{\Delta G^{\circ}_{318}}{-nF} = \frac{(-201.1\ kJ)\left(\frac{10^{3} J}{kJ}\right)}{-(2)\ (96{,}500\ C)\left(\frac{J}{C \times V}\right)} = 1.04\ V$$

$$E_{318} = E^{\circ}_{318} - \frac{(2.303)(8.314\ J/mol{\bullet}K)(318\ K)}{(2)(96{,}500\ J/V{\bullet}mol)} \log \frac{(0.100)}{(0.0500)^{2}}$$

$E = 1.04\ V - 0.05\ V = \mathbf{0.99\ V}$

19 CHEMICAL KINETICS: THE STUDY OF THE RATES OF CHEMICAL REACTIONS

19.1 (a) the nature of reactants and products (b) concentration of reacting species (c) temperature (d) influence of outside agents (catalysts)

19.3 Methods used to study the rate of a reaction must be fast, accurate and not interfere with the normal course of the reaction.

19.5 We say that reaction rate depends upon the nature of the reactants because some substances react rapidly with one another by the nature of the reaction taking place while others react more slowly. In other words, even under conditions of equal concentrations and temperature, different chemical reactions progress at different rates.

19.7 Reaction rate = the speed at which reactants are consumed or the products are formed. It is the ratio of the change in concentration to the change in time units, such as mol $liter^{-1}$ s^{-1}.

19.9 $CH_4 + 2O_2 \rightarrow CO_2 + 2H_2O$

$$\text{rate (for } CO_2) = \frac{0.16 \text{ mol } CH_4}{L \times s} \times \frac{1 \text{ mol } CO_2}{1 \text{ mol } CH_4} = \mathbf{0.16 \text{ mol } CO_2 \ L^{-1} \ s^{-1}}$$

$$\text{rate (for } H_2O) = \frac{0.16 \text{ mol } CH_4}{L \times s} \times \frac{2 \text{ mol } H_2O}{1 \text{ mol } CH_4} = \mathbf{0.32 \text{ mol } H_2O \ L^{-1} \ s^{-1}}$$

19.11 $2A \rightarrow 4B + C$

From your graph, slope of the disappearance of A at 25 min =

$$\frac{\Delta[A]}{\Delta t} \approx \mathbf{-9.6 \times 10^{-3} \text{ mol A } L^{-1} \text{ min}^{-1}}$$

The slope for the rate of formation B at 25 min =

$$\frac{\Delta[B]}{\Delta t} \approx \mathbf{1.9 \times 10^{-2} \text{ mol B } L^{-1} \text{ min}^{-1}}$$

The slope of the disappearance of A at 40 min =

$$\frac{\Delta[A]}{\Delta t} \approx \mathbf{-6.4 \times 10^{-3} \text{ mol A } L^{-1} \text{ min}^{-1}}$$

The slope of the formation of B at 40 min =

$$\frac{\Delta[B]}{\Delta t} \approx \mathbf{1.3 \times 10^{-2} \text{ mol B } L^{-1} \text{ min}^{-1}}$$

Rate of B = -2 x rate of A

$\frac{\Delta[C]}{\Delta t}$ at 25 min $\approx$ $\mathbf{4.8 \times 10^{-3} \text{ mol C mol}^{-1} \ L^{-1}}$

$\frac{\Delta[C]}{\Delta t}$ at 40 min $\approx$ $\mathbf{3.2 \times 10^{-3} \text{ mol C } L^{-1} \text{ min}^{-1}}$

19.13 The sum of exponents on the concentrations in the rate law is the order of a reaction.

19.15 As the concentration of CO varies, the rate at which it is removed from the earth's atmosphere by fungi remains constant (no concentration dependence); therefore, it appears to be a zero-order reaction.

19.17 (a) liter/mol s (b) liter/mol s (c) $liter^3/mol^3$ s

19.19 -1; i.e., Rate = $k[A]^{-1}$

19.21 (a) rate = $(2.35 \times 10^{-6} L^2 mol^{-2} s^{-1})(1.00 mol L^{-1})^2(1.00 mol L^{-1})$

$= \mathbf{2.35 \times 10^{-6} mol L^{-1} s^{-1}}$

(b) rate = $(2.35 \times 10^{-6} L^2 mol^{-2} s^{-1})(0.250 mol L^{-1})^2(1.30 mol L^{-1})$

$= \mathbf{1.91 \times 10^{-7} mol L^{-1} s^{-1}}$

19.23 (a) rate = $k[NO_2]^x[O_3]^y$ In the first and second experiments, the initial NO_2 concentration is constant, the initial concentration of O_3 is doubled, and the rate is doubled. Therefore, the value of y must be 1. In the second and third experiments, the initial concentration of O_3 is constant, the initial concentration of NO_2 is halved, and the rate is halved. Therefore, the value of x must be 1. **Rate = $k[NO_2][O_3]$**

(b) **k** = Rate/$[NO_2][O_3]$ = 0.022 mol $L^{-1} s^{-1}$/
$[(5.0 \times 10^{-5} mol L^{-1})(1.0 \times 10^{-5} mol L^{-1})]$ = $\mathbf{4.4 \times 10^7 L mol^{-1} s^{-1}}$

19.25 (a) Rate = $k[NOCl]^x$ When the concentration was doubled, the rate changed by a factor of $1.44 \times 10^{-8}/3.60 \times 10^{-9}$ or 4.00. Therefore, x must be 2.
Rate = $k[NOCl]^2$

(b) **k** = Rate/$[NOCl]^2$ = 3.60×10^{-9} mol $L^{-1} s^{-1}/(0.30 mol L^{-1})^2$
$= \mathbf{4.0 \times 10^{-8} L mol^{-1} s^{-1}}$

(c) $(0.45/0.30)^2$ = **2.2 times faster**

19.27 The half-life of a reaction is the time required for the concentration of a given reactant to be decreased by a factor of 2 (i.e., to half of its initial value).

19.29 See Figure 19.4, except each $t_{1/2}$ is twice the length of the preceding $t_{1/2}$.

19.31 $t_{1/2} = \dfrac{1}{k[B]_0}$

$$2.11 \text{ min} = \frac{1}{k(0.10 \text{ M})}$$

$$k = 1/[(0.10 \text{ M})\ (2.11 \text{ min})]$$

$$k = 4.7 \text{ M}^{-1} \text{ min}^{-1}$$

$$t_{1/2} = \frac{1}{(4.7 \text{ M}^{-1} \text{ min}^{-1})\ (0.010 \text{ M})} = \mathbf{21\ min}$$

19.33 (a) $t_{1/2} = 0.693/k = 0.693/3.2 \times 10^{-2} \text{ s}^{-1} = \mathbf{22\ s}$

(b) $\ln \dfrac{[A]_0}{[A]_t} = kt \qquad \ln \dfrac{[A]_0}{(0.010 \text{ M})} = (3.2 \times 10^{-2} \text{ s}^{-1})\ (60 \text{ s}) = 1.92$

$$\frac{[A]_0}{(0.010 \text{ M})} = 6.8 \qquad [A]_0 = 6.8 \times 10^{-2} \text{ M or } \mathbf{0.068\ M}$$

19.35 An overall chemical reaction represents the net chemical change. This does not mean that all the reactants come together simultaneously. To predict the rate law, a reaction mechanism must be known. From the mechanism, reaction order is known. The reaction order can also be obtained experimentally and from that a rate law can be written.

19.37 A one-step mechanism would involve the simultaneous collision of six molecules, five of which would have to be O_2. This is very improbable.

19.39 The rate law is Rate = $k[NO_2]^2$. CO does not appear in the rate-determining step (slow step), and, therefore, CO does not affect the rate and the reaction is said to be zero-order with respect to CO.

19.41 (a) $2A + B \rightarrow C + 2D$

(b) Rate = $k[A]^2$

(c) Rate = $k[A]^2[B]$

19.43 The observed rate of reaction is much smaller (for example, approximately 5 x 10^{12} for $2HI(g) \rightarrow H_2(g) + I_2(g)$) than what is expected if all the collisions were effective. A minimum amount of energy is required to cause a reaction to occur, and the molecules must collide with the proper orientation.

19.45

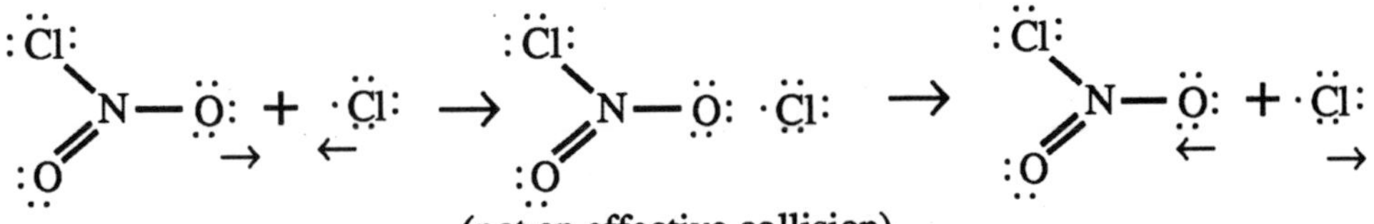

:Cl: / :O= N—O: + ·Cl: → :Cl: / :O= N—O: ·Cl: → :Cl: / :O= N—O: + ·Cl:

(not an effective collision)

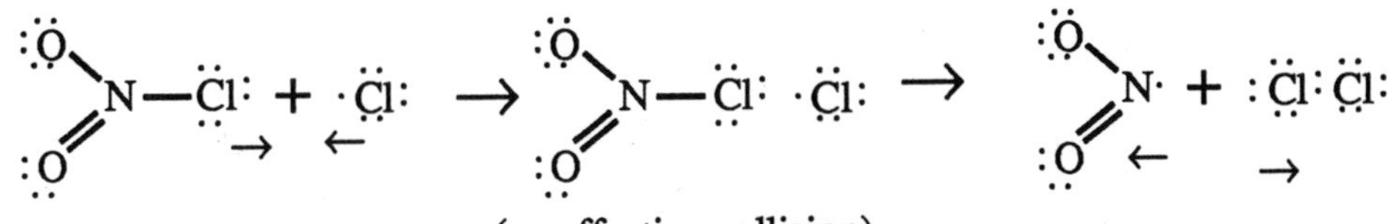

:O / :O= N—Cl: + ·Cl: → :O / :O= N—Cl: ·Cl: → :O / :O= N· + :Cl:Cl:

(an effective collision)

19.47 The **activation energy** represents the kinetic energy required to bring the reactants to the point where they can react to form products. It is the minimum kinetic energy that must be available in a collision.

19.49 Reactions, including biochemical ones, slow down at low temperatures primarily because a smaller fraction of molecules possesses the required activation energy.

19.51 The activated complex is the intermediate species that is highly unstable, transient and very reactive. The activated complex exists in a transition state between products and reactants. The transition state is on the barrier (high potential energy) of the potential energy curve.

19.53 (See Problem 19.52)

$$\ln\left(\frac{k_1}{k_2}\right) = \frac{E_a}{R}\left(\frac{1}{T_2} - \frac{1}{T_1}\right)$$

$\ln (1.32 \times 10^{-2}/1.64) = (E_a/8.314 \text{ J mol}^{-1}\text{K}^{-1})[(1/548 \text{ K}) - (1/473 \text{ K})]$

$-4.822 = (E_a/8.314 \text{ J mol}^{-1} \text{ K}^{-1})(0.001825 - 0.002114)$

$E_a = (-4.822)(8.314 \text{ J mol})/(-0.000289)$

$\mathbf{E_a} = 1.387 \times 10^5 \text{ J mol}^{-1} = \mathbf{139\ kJ\ mol^{-1}}$

$$k = Ae^{-E_a/RT} \qquad A = \frac{k}{e^{-E_a/RT}} = ke^{E_a/RT}$$

$A = (1.32 \times 10^{-2} \text{ L mol}^{-1} \text{ s}^{-1})\ e^{1.39 \times 10^5 \text{ J mol}^{-1}/(8.314 \text{ J mol}^{-1} \text{ K}^{-1})(473 \text{ K})}$

$A = (1.32 \times 10^{-2} \text{ L mol}^{-1} \text{ s}^{-1})(e^{35.3}) = \mathbf{3 \times 10^{13}\ L\ mol^{-1}\ s^{-1}}$

19.55 $\ln\left(\frac{k_1}{k_2}\right) = \frac{E_a}{R}\left(\frac{1}{T_2} - \frac{1}{T_1}\right)$

$$\ln\left(\frac{1.32 \times 10^{-2} \text{ L mol}^{-1} \text{ s}^{-1}}{k_2}\right) = \frac{138 \times 10^3 \text{ J mol}^{-1}}{8.314 \text{ J mol}^{-1} \text{ K}^{-1}} [(1/573 \text{ K}) - (1/473 \text{ K})]$$

$\ln (1.32 \times 10^{-2} \text{ L mol}^{-1} \text{ s}^{-1}/k_2) = -6.12$

$$\frac{1.32 \times 10^{-2} \text{ L mol}^{-1} \text{ s}^{-1}}{k_2} = e^{-6.12} = 0.0022$$

$\mathbf{k_2 = 6.0 \text{ L mol}^{-1} \text{ s}^{-1}}$

19.57 $\ln\left(\frac{k_1}{k_2}\right) = \frac{E_a}{R}\left(\frac{1}{T_2} - \frac{1}{T_1}\right)$

$$\ln \frac{(3.2 \times 10^{-2})}{(9.3 \times 10^{-2})} = \frac{E_a}{8.314 \text{ J mol}^{-1} \text{ K}^{-1}} [(1/848 \text{ K}) - (1/823 \text{ K})]$$

$$\frac{(-1.1)(8.314 \text{ J mol}^{-1})}{(0.001179 - 0.001215)} = E_a = 2.48 \times 10^5 \text{ J mol}^{-1} = \mathbf{250 \text{ kJ mol}^{-1}}$$

19.59 (See Problem 19.58)

$\ln k = \ln A - \frac{E_a}{R}\left(\frac{1}{T}\right)$ k is proportional to 1/time, therefore:

$\ln \frac{1}{t} = \ln A - \frac{E_a}{R}\left(\frac{1}{T}\right)$

Plot ln (1/t) vs. 1/T (T = kelvins)

(continued)

19.59 (continued)

$$\text{Slope} = -\frac{E_a}{R}$$

$\mathbf{E_a = 64\ kJ\ mol^{-1}}$

Estimated by extrapolation, when $1/T = 1/288 = 3.47 \times 10^{-3}$, $\ln 1/t = -2.64$ or

$\ln t = 2.64$.

Therefore, it was estimated that **time = ~13 min.**

19.61 (a) A **heterogeneous catalyst** is a substance that provides a low energy pathway (E_a) to the products, but is not in the same phase as the reactants. These catalysts appear to adsorb reactant molecules and certain bonds within the reactants are weakened or broken.

(b) An **inhibitor** interferes with the effectiveness of a catalyst by interfering with adsorption.

19.63 The catalyst lowers the activation energy by giving the reactants a different pathway (mechanism) for the chemical reaction. This path has a lower E_a, thereby increasing the number of effective collisions.

19.65 Free radicals are extremely reactive intermediate substances consisting of atoms or groups of atoms that possess unpaired electrons. They are formed either thermally or by absorption of photons of appropriate frequencies.

19.67 Free radicals are dangerous in living organisms because they are so reactive and may cause aging and cancer.

19.69 (a) initiation step 1

(b) propagation steps 3, 5, 6

(c) termination step 2

19.71 t = # + 4 Where # = no. of chirps in 8 seconds

(a)

20 = # + 4	no. of chirps = **16** at 20°C
25 = # + 4	no. of chirps = **21** at 25°C
30 = # + 4	no. of chirps = **26** at 30°C
35 = # + 4	no. of chirps = **31** at 35°C

(b) E_a = slope x (-R) = (-3.9 x 10^3) x (-8.314 x 10^{-3} kJ/moles) = **32kJ/mole**

(c) t(°C) = # + 4; 120°C = # + 4
# of chirps = **116** (From the graph, a somewhat smaller value was calculated. However, in reality, the value is probably very small since the cricket is unlikely to survive 8 sec at 120°C.)

19.73 $2O_3 + h\nu \rightarrow 3O_2$
Since this is a chain reaction, small amounts of intermediate can bring about the destruction of very large amounts of ozone.

20 METALS AND THEIR COMPOUNDS; THE REPRESENTATIVE METALS

20.1 Most metals are found in the combined state. Sources of metals are the ocean and land-based deposits of metal carbonates, sulfates, oxides, and sulfides.

20.3 The three steps are: concentration, reduction, and refining.

20.5 An amalgam is a solution of a metal in mercury. Gold ore is mixed with mercury, which dissolves the metallic gold. The mercury is then separated from the stone and distilled, leaving the pure gold behind.

20.7 (a) $Al_2O_3 + 2OH^- \rightarrow 2AlO_2^- + H_2O$

(b) $AlO_2^- + H_2O + H^+ \rightarrow Al(OH)_3$

(c) $2Al(OH)_3 \xrightarrow{\text{heat}} Al_2O_3 + 3H_2O$

20.9 It must have a small or negative ΔH°_f so that the reaction proceeds to completion at a reasonable temperature, i.e., that ΔG°_T is negative.

20.11 $2Ag_2O \rightarrow 4Ag + O_2$

ΔH (reaction) = [(4 x 0) + (1 x 0)] - [2 mol x (-30.5 kJ/mol)] = 61.0 kJ

ΔS (reaction) = -2 mol x ΔS_f = -2 mol x (-66.1 J/mol K) = 132.2 J/K

$\Delta G = \Delta H - T\Delta S$ $\quad\quad$ $\Delta G = -RT \ln K$ $\quad$ (if K = 1, then $\Delta G = 0$)

When K = 1, $0 = \Delta H - T\Delta S$ $\quad\quad$ 0 = 61.0 kJ - T(132.2 J/K)

T = (61.0 x 10^3/132.2)K = **461 K** $\quad\quad$ **188°C**

20.13 $ZnO(s) \rightarrow Zn(s) + 1/2\ O_2(g)$ $\quad\quad$ $\Delta H^\circ = +348$ kJ

$\Delta S^\circ = (41.8 + 1/2 \times 205.0 - 43.5)\ J\ K^{-1} = 100.8\ J\ K^{-1}$

$\Delta G^\circ = \Delta H^\circ - T\Delta S^\circ = 348\ kJ - T(100.8 \times 10^{-3}\ kJ\ K^{-1}) = 0$

T = 3450 K (or **3180°C**)

20.15 $\Delta G^\circ = -RT \ln K_p$ and $\Delta G^\circ = \Delta H^\circ - T\Delta S^\circ$; $\quad$ $\Delta G^\circ = 754.4\ kJ - T\left(\frac{257.9\ J}{K}\right)$

$\Delta G^\circ_{373} = 658.2$ kJ, $\Delta G^\circ_{773} = 555$ kJ, $\Delta G^\circ_{2273} = 168.2$ kJ

$\ln K_p = -\Delta G^\circ/RT$ $\quad\quad$ $K_p = e^{-\Delta G^\circ/RT}$

$\mathbf{K_{p(373)} = 7 \times 10^{-93}}$

$\mathbf{K_{p(773)} = 3 \times 10^{-38}}$

$\mathbf{K_{p(2273)} = 1.36 \times 10^{-4}}$

20.17 It is plentiful and least expensive.

20.19 The chemical reducing agent itself would be even more difficult to prepare.

20.21 $2NaCl(\ell) \xrightarrow{\text{electrolysis}} 2Na(\ell) + Cl_2(g)$

$$2Al_2O_3(\ell) \xrightarrow[\text{cryolite}]{\text{electrolysis}} 4Al(\ell) + 3O_2(g)$$

20.23 The Mond process is used in the refining of nickel. In this process impure nickel is treated with carbon monoxide at moderately low temperatures. The $Ni(CO)_4$ gas that is produced is then separated and heated to 200°C and decomposes to give pure nickel plus CO.

20.25 (a) Li (b) Al (c) Cs (d) Sn (e) Ga

20.27 Al_2O_3

20.29 Many are amphoteric and form compounds with some covalent properties.

20.31 (a) $GeCl_4$ (b) Bi_2O_5 (c) PbS (d) Li_2S (e) MgS

20.33 Charge transfer from anion to cation which absorbs photons in the visible portion of the spectrum is the phenomenon responsible for the color of compounds such as SnS_2 and PbS.

20.35 Red-violet

20.37 They are relatively rare and, therefore, more expensive to produce. Compounds of Na and K usually serve just as well.

20.39 $Na^+(g) \rightarrow Na^+(aq) \quad \Delta H^\circ = ?$

(1) $Na(s) \rightarrow Na^+(aq) + e^- \quad \Delta H^\circ_1 = -239.7$

(2) $Na(g) \rightarrow Na(s) \quad \Delta H^\circ_2 = -108.7$

(3) $Na^+(g) + e^- \rightarrow Na(g) \quad \Delta H^\circ_3 = -493.7$

$\Delta H^\circ = \Delta H^\circ_1 + \Delta H^\circ_2 + \Delta H^\circ_3 =$ **-842.1 kJ mol^{-1}**

20.41

K(s)	→	$K^+(aq)$	+ 1e$^-$	$\Delta H^\circ = ?$
K(s)	→	K(g)		$\Delta H^\circ = 90.0$ kJ/mol
K(g)	→	$K^+(g)$	+ 1e$^-$	$\Delta H^\circ = 418$ kJ/mol
$K^+(g)$	→	$K^+(aq)$		$\Delta H^\circ = -759$ kJ/mol
				$\Delta H^\circ = -251$ kJ/mol

$\Delta G^\circ = -nFE^\circ = \Delta H^\circ - T\Delta S^\circ$

$$\Delta S^\circ = \frac{-nFE^\circ - \Delta H^\circ}{-T} = \frac{[1 \times 96{,}500 \text{ C/mol} \times 2.92 \text{ V} \times \text{J/(V} \times \text{C)}]}{298 \text{ K}} + \frac{(-251 \times 10^3 \text{ J/mol})}{298 \text{ K}}$$

$$= \mathbf{103\ J\ K^{-1}\ mol^{-1}}$$

20.43 Alkali metals occur as compounds in the ocean and in salt deposits.

20.45 Use of sodium in the cooling of nuclear reactors takes advantage of its low melting point, relatively high boiling point, and good thermal conductivity. Using sodium in vapor lamps takes advantage of sodium's emission spectra.

20.47 To detect potassium in the presence of sodium, view their flame through blue "cobalt glass" to filter out the yellow light from sodium.

20.49 Because of the very large hydration energy of the tiny Li^+ ion.

20.51 (a) $2Li + Br_2 \rightarrow 2LiBr$

$2Na + Br_2 \rightarrow 2NaBr$

(b) $2Li + S \rightarrow Li_2S$

$2Na + S \rightarrow Na_2S$

(c) $6Li + N_2 \rightarrow 2Li_3N$

$Na + N_2 \rightarrow$ no reaction

20.53 KO_2 is used in a recirculating breathing apparatus.

$4KO_2(s) + 2CO_2(g) \rightarrow 2K_2CO_3(s) + 3O_2(g)$

$2KO_2 + 2H_2O \rightarrow 2KOH + O_2 + H_2O_2$

20.55 A metal that in the production of electronic vacuum tubes reacts with traces of O_2 and H_2O and removes them from an otherwise inert atmosphere is sometimes referred to as a getter.

20.57 Trona ore is: $Na_2CO_3 \cdot NaHCO_3 \cdot 2H_2O$

Solvay process:

$CaCO_3 \xrightarrow{heat} CaO + CO_2$

$CO_2 + H_2O \rightarrow H_2CO_3$

$H_2CO_3 + NH_3 \rightarrow NH_4^+ + HCO_3^-$

$HCO_3^- + Na^+ + Cl^- \rightarrow NaHCO_3 + Cl^-$

$2NaHCO_3 \xrightarrow{heat} Na_2CO_3 + H_2O + CO_2$

$CaO + H_2O \rightarrow Ca(OH)_2$

$2NH_4Cl + Ca(OH)_2 \rightarrow CaCl_2 + 2NH_3 + 2H_2O$

Net reaction: $2NaCl + CaCO_3 \rightarrow Na_2CO_3 + CaCl_2$

20.59 Potash is K_2CO_3.

It is made in the following way:

$KOH + CO_2 \rightarrow KHCO_3$

$2KHCO_3 \xrightarrow{heat} K_2CO_3 + H_2O + CO_2$

20.61 Group IIA elements are called alkaline earth metals because their oxides are basic and their ores are found in the earth. Their densities, hardness, melting points and ionization energies are all greater than those of Group IA metals.

20.63 From dolomite:

$$CaCO_3 \cdot MgCO_3 \xrightarrow{\text{heat}} CaO \cdot MgO + 2CO_2$$
$$CaO + H_2O \rightarrow Ca^{2+} + 2OH^-$$
$$MgO + H_2O \rightarrow Mg(OH)_2(s)$$
$$Mg(OH)_2(s) + 2HCl \rightarrow MgCl_2 + 2H_2O$$
$$MgCl_2(\ell) \xrightarrow{\text{electrolysis}} Mg(\ell) + Cl_2(g)$$

From sea water:

$$CaO + H_2O + Mg^{2+} \rightarrow Ca^{2+} + Mg(OH)_2(s)$$
$$Mg(OH)_2(s) + 2HCl \rightarrow MgCl_2 + 2H_2O$$
$$MgCl_2(\ell) \xrightarrow{\text{electrolysis}} Mg(\ell) + Cl_2(g)$$

20.65 Calcining: heating a substance strongly decomposes carbonate to the oxide and CO_2.
Lime is CaO.
$$CaO + H_2O \rightarrow Ca(OH)_2$$
Lime is an inexpensive, relatively strong base.

20.67 (a) brick-red (b) crimson (c) yellowish-green

20.69
$$2Mg + O_2 \rightarrow 2MgO$$
$$Mg + S \rightarrow MgS$$
$$3Mg + N_2 \rightarrow Mg_3N_2$$

20.71 Covalently linked chain of $BeCl_2$ units. See Figure 20.18 and the drawing in the text beside Section 20.6. Formation of the additional two Be—Cl coordinate covalent bonds suggests that the Be in $BeCl_2$ seeks additional electrons to complete its octet.

20.73 (a) $Ca + 2H_2O \rightarrow Ca(OH)_2 + H_2$

(b) $2K + 2H_2O \rightarrow 2KOH + H_2$

20.75 Calcium carbonate is used to make lime, as a mild abrasive, as an antacid, and as the main ingredient in chalk.

20.77 Gypsum is: $CaSO_4 \cdot 2H_2O$

Making of plaster of Paris: $CaSO_4 \cdot 2H_2O \xrightarrow{heat} CaSO_4 \cdot 1/2H_2O + 3/2H_2O$

Reaction of plaster of Paris with water: $CaSO_4 \cdot 1/2H_2O + 3/2H_2O \rightarrow CaSO_4 \cdot 2H_2O$

20.79 Epsom salts is: $MgSO_4 \cdot 7H_2O$. MgO is used in refractory bricks, the manufacture of paper, and, medicinally, as an antacid.

20.81 The post-transition metals are Ga, In, Tl, Sn, Pb and Bi.
Lower oxidation states become more stable going down a group because the energy needed to remove additional electrons is less likely to be recovered by forming additional bonds since bond strengths decrease as the atoms become larger.

20.83 Aluminum is used as a structural metal, in kitchen utensils, automobiles, aircraft, beverage cans, aluminum foil, other consumer products, electrical wiring and in alnico for magnets.

20.85 $2Al(s) + 6H^+(aq) \rightarrow 2Al^{3+}(aq) + 3H_2(g)$

$2Al(s) + 2OH^-(aq) + 2H_2O(\ell) \rightarrow 2AlO_2^-(aq) + 3H_2(g)$

20.87 The thermite reaction is: $Fe_2O_3(s) + 2Al(s) \rightarrow Al_2O_3(\ell) + 2Fe(\ell) + heat$

20.89 Solutions of aluminum salts are acidic because hydrolysis will produce hydrated protons:
$Al(H_2O)_6^{3+} + H_2O \rightleftharpoons [Al(H_2O)_5OH]^{2+} + H_3O^+$

20.91 The aluminate ion may be written as either: AlO_2^- or $Al(H_2O)_2(OH)_4^-$

20.93 An alum is a double salt of the general formula $M^+M^{3+}(SO_4)_2 \cdot 12H_2O$ ($M^+ = Na^+$, NH_4^+, or K^+ and $M^{3+} = Al^{3+}$, Cr^{3+}, or Fe^{3+}). An example is $NaAl(SO_4)_2 \cdot 12H_2O$. This is the alum in baking powders. It evolves CO_2 from $NaHCO_3$ because the aluminum ion hydrolyzes releasing H_3O^+ which reacts with the $NaHCO_3$.

20.95 Tin from its ore, cassiterite: $SnO_2 + C \rightarrow Sn + CO_2$

Lead from its ore, galena: $2PbS + 3O_2 \rightarrow 2PbO + 2SO_2$
$2PbO + C \rightarrow 2Pb + CO_2$

Bismuth from its ore: $2Bi_2S_3 + 9O_2 \rightarrow 2Bi_2O_3 + 6SO_2$
$2Bi_2O_3 + 3C \rightarrow 4Bi + 3CO_2$

20.97 Allotropes are different physical forms of the same element. Tin exhibits allotropism.

20.99 $Sn + 4HNO_3 \rightarrow SnO_2 + 4NO_2 + 2H_2O$
$3Pb + 8HNO_3 \rightarrow 3Pb(NO_3)_2 + 2NO + 4H_2O$
$Sn + 2Cl_2 \rightarrow SnCl_4$
$Pb + Cl_2 \rightarrow PbCl_2$
In each case, tin yields the 4+ ion; lead yields the 2+ ion.

20.101 $Sn(s) + 2OH^-(aq) + 2H_2O(\ell) \rightarrow Sn(OH)_4^{2-}(aq) + H_2(g)$

$Pb(s) + 2OH^-(aq) + 2H_2O(\ell) \rightarrow Pb(OH)_4^{2-}(aq) + H_2(g)$

20.103 Lead based paints slowly darken due to the reaction of Pb^{2+} with airborne H_2S to give black PbS.

21 THE TRANSITION METALS

21.1 A transition element is one that possesses a partially filled or filled d subshell and fits between Groups IIA and IIIA.

21.3 Compounds of elements in the corresponding A and B groups have similar composition, structure, and maximum positive oxidation states.

21.5 Fe, Co, Ni

21.7 Four general properties of the transition elements are:
(1) They exhibit multiple oxidation states.
(2) Many transition metal compounds are paramagnetic.
(3) Many of their compounds are colored.
(4) They tend to form complex ions.

21.9 $KMnO_4$

21.11 Most have a pair of electrons in an s orbital in the shell with the largest value of the principal quantum number. These are the first and easiest electrons to be lost during ionization; therefore, a common oxidation state is the +2 in which only the s electrons have been removed.

21.13 Going from left to right in a period, the lower oxidation states become relatively more stable. Going from top to bottom in a d-block group, the higher oxidation states become relatively more stable.

21.15 Ni^{3+}

21.17 Cu^{3+}

21.19 The differences in electron configuration occur in a subshell that is two shells below the outer shell. Therefore, many chemical properties of the lanthanide elements are very similar.

21.21 They have the same electron structure in their outer shells and because of the lanthanide contraction, Hf is very nearly the same size as Zr. Thus, their chemical properties are very similar.

21.23 The effects of increased nuclear charge by the addition of protons and increased shielding by the electrons added to the d subshell just beneath the outer shell nearly off-set each other in the transition elements.

21.25 Paramagnetic substances are weakly drawn into a magnetic field, whereas, ferromagnetic substances are strongly drawn ($\sim 10^6$ times stronger than paramagnetic substances) into a magnetic field. Both properties are a result of unpaired electrons. A ferromagnetic substance can become permanently magnetized by placing it in a strong magnetic field so that the domains become aligned with the field. The domains remain aligned even after removal of the external field.

21.27 Sc, Y, La and lanthanide elements (atomic numbers 58 to 71)

$$2M + 6H_2O \rightarrow 2M(OH)_3 + 3H_2$$

21.29 Oxides of metals in high oxidation states tend to be acidic anhydrides since when they hydrate they tend to have one or more non-hydrated oxygens that, along with an increased charge of the central atom strengthens the X-O bond and weakens the O-H bond.

21.31 The energy for removal of 4 electrons is so high that Ti^{+4} ion does not have a real existence.

21.33 Chromium is used to coat other metals because it is lustrous and very resistant to corrosion. The most important oxidation states of chromium are 0, 3+, and 6+.

21.35 Cr(III) in water is acidic because it forms $Cr(H_2O)_6^{3+}$ which is a weak acid and undergoes the following reaction with base: $Cr(H_2O)_6^{3+}(aq) + 3OH^-(aq) \rightarrow Cr(H_2O)_3(OH)_3(s) + 3H_2O(\ell)$

21.37 See the drawing at the end of the section on chromium in the text.

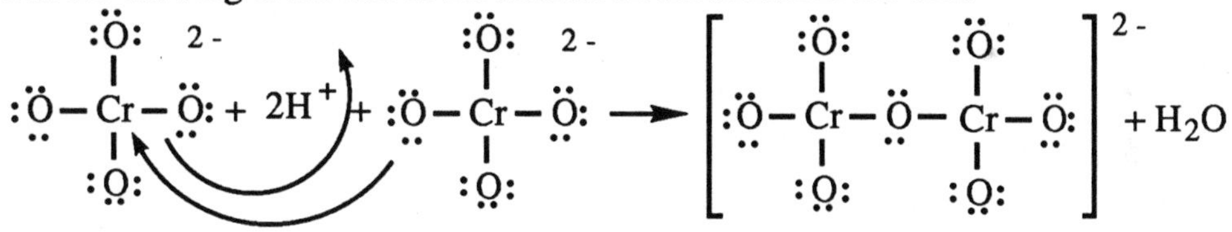

$$[O-Cr(O)_2-O]^{2-} + 2H^+ + [O-Cr(O)_2-O]^{2-} \longrightarrow [O-Cr(O)_2-O-Cr(O)_2-O]^{2-} + H_2O$$

21.39 $KMnO_4$ is a deeply colored strong oxidizing agent and its reduction product in acidic solution, Mn^{2+}, is nearly colorless. In neutral or basic solution, the reduction product is brown, insoluble MnO_2 which obscures the endpoint. Therefore, $KMnO_4$ is a useful titrant only in acidic solutions.

21.41 $3MnO_4^{2-} + 4H^+ \rightarrow 2MnO_4^- + MnO_2 + 2H_2O$

21.43 FeO, Fe_2O_3, Fe_3O_4. Only Fe_3O_4 is magnetic.

21.45 When Fe^{3+} and $Fe(CN)_4^{4-}$ are mixed, $Fe_4[Fe(CN)_6]_3 \cdot 16H_2O$ is formed. This product is known as Prussian blue.

21.47 Cobalt is used in many alloys with special properties such as in high-temperature alloys that are employed in tools for cutting and machining other metals at high speed and in catalysts. The principal oxidation states of cobalt are +2 and +3.

21.49 Solutions of nickel salts are green because the ion usually formed in water, $Ni(H_2O)_6^{2+}$, is green. NiO_2 is used as the cathode in nickel-cadmium batteries.

21.51 Copper is used in electrical wire. Silver is used in photography. Gold is used in the plating of low voltage electrical contacts.

21.53 H^+ is not a strong enough oxidizing agent to dissolve the coinage metals. Copper and silver react with dilute HNO_3.

$3Cu + 8HNO_3 \rightarrow 3Cu(NO_3)_2 + 2NO + 4H_2O$

$(3Cu(s) + 8H^+(aq) + 2NO_3^-(aq) \rightarrow 3Cu^{2+}(aq) + 2NO(g) + 4H_2O(\ell))$

$3Ag + 4HNO_3 \rightarrow 3AgNO_3 + NO + 2H_2O$

$(3Ag(s) + 4H^+(aq) + NO_3^-(aq) \rightarrow 3Ag^+(aq) + NO(g) + 2H_2O(\ell))$

21.55 AgCl, AgBr and AgI are used in photographic films and papers.

21.57 The deep blue $Cu(NH_3)_4^{2+}$ ion is formed when ammonia is added to an aqueous solution of pale blue $Cu(H_2O)_4^{2+}$.

21.59 A test for superconductivity is to see if the substance is repelled by a magnet after being cooled to the necessary temperature.

21.61 The protection of a metal from corrosion by being in contact with a metal that is more easily oxidized and which will be preferentially oxidized is known as cathodic protection.

21.63 Cadmium is used in place of zinc when a basic environment is anticipated since zinc is attacked by base but cadmium is not. Cadmium is not used more often because it is less abundant than zinc (therefore more expensive) and because its salts are very toxic.

21.65 Common uses of zinc oxide include paint pigment, sun screen, and fast-setting dental cements.

21.67 To test for Hg_2^{2+} add HCl, which will cause Hg_2Cl_2 to precipitate. Treatment of the Hg_2Cl_2 with aqueous NH_3 will give a black precipitate because of disproportionation to Hg(ℓ) and $Hg(NH_2)Cl(s)$.

21.69 $K_{sp} = [Hg^+]\ [Cl^-] = 2.2 \times 10^{-18}$, 4.4×10^{-18}, 1.1×10^{-17}, and 2.2×10^{-17}

Not Constant

$K_{sp} = [Hg_2^{2+}]\ [Cl^-]^2 = 1.1 \times 10^{-18}$, 1.1×10^{-18}, 1.1×10^{-18}, and 1.1×10^{-18}

Constant Value **correct formula = Hg_2^{2+}**

21.71 To test for cyanide, one could add Fe^{2+}, then Fe^{3+} to form the deep blue precipitate $Fe_4[Fe(CN)_6]_3 \cdot 16H_2O$.

21.73 Also see Section 10.7 in the textbook.

(a) Oxalate

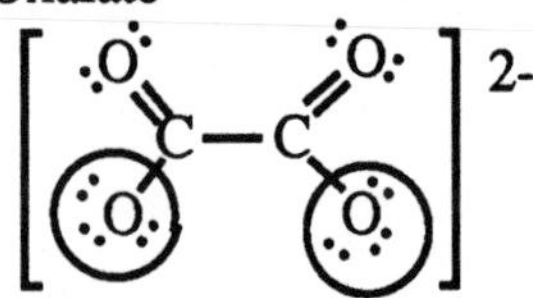

(b) Ethylenediamine

H_2C—CH_2

H_2N NH_2

(c) Ethylenediaminetetraacetate ion

:O:

:O—C—H_2C CH_2 C—O:

:N—CH_2—CH_2—N:

:O—C—H_2C CH_2 C—O:

:O: :O:

21.75 See Figure 21.16 for assistance in drawing the linear, tetrahedral, square planar, and octahedral structures requested.

21.77 NTA can coordinate to four sites in an octahedral complex in much the same manner as EDTA.

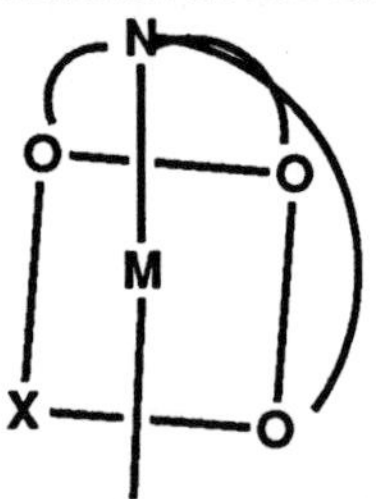

The remaining two sites (X) can be occupied by other ligands or by H_2O molecules. The NTA would increase the solubility of metal salts by shifting to the right equilibria such as: $MX_n(s) + NTA \rightleftharpoons M(NTA)X_{n-4}\,(aq) + 4X(aq)$

21.79 Isomers of $[Co(NH_3)_2Cl_4]^-$

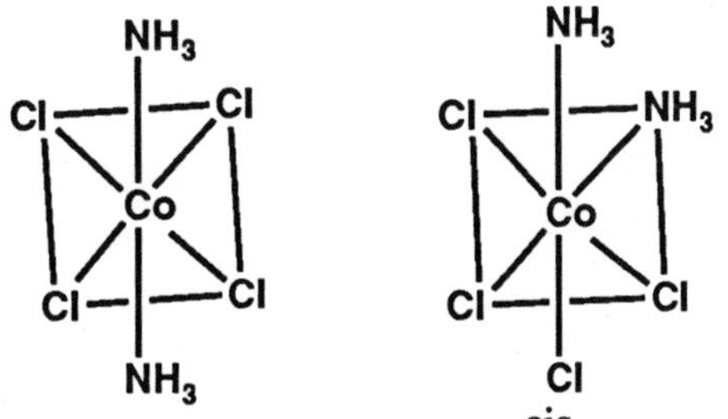

trans cis

Isomers of $[Co(NH_3)_3Cl_3]$

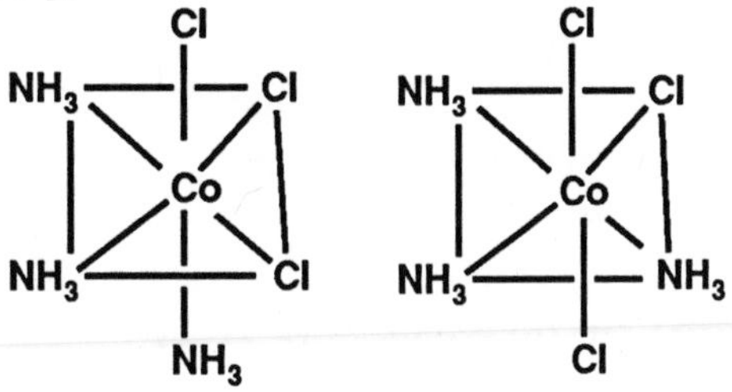

21.81 Chiral substances are said to be optically active because they or their solutions rotate the plane of plane-polarized light when plane-polarized light is passed through them or their solutions.

21.83

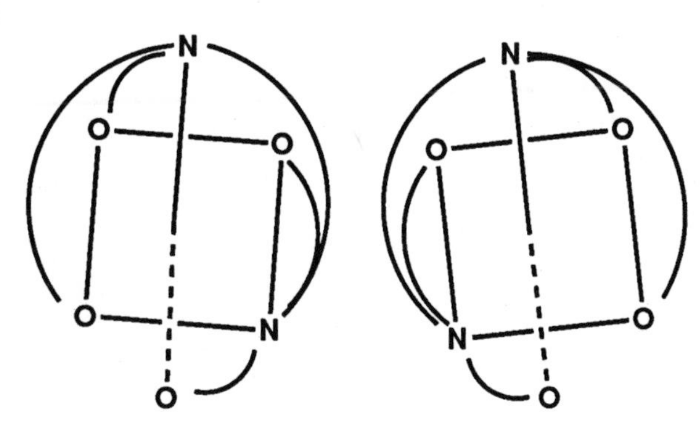

21.85 (a) four (high spin)

(b) two (low spin)

21.87

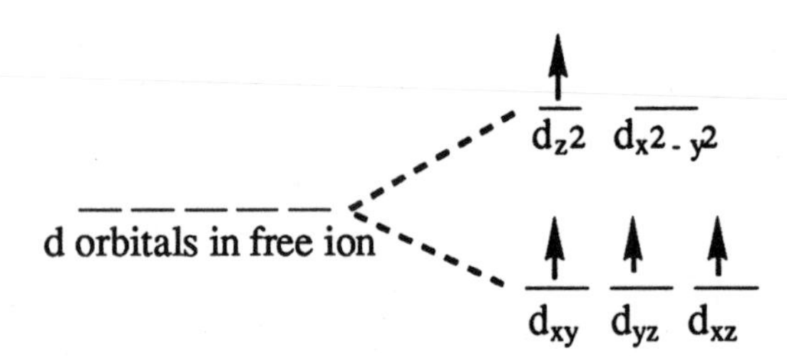

21.89 If $\Delta < P$,the electrons occupy all orbitals before pairing and a paramagnetic complex will be formed unless the orbital set is filled; when $\Delta > P$, the electrons pair up in the lower energy levels before occupying the higher energy levels; this increases the possibility of a diamagnetic complex (See Question 21.92 c).

21.91 (a) See Figure 21.33 (b) See Figure 21.35

21.93 (a) 2 d-electrons -- weak ligand -- 2 unpaired electrons -- inner complex
(b) 8 d-electrons -- moderate ligand -- 2 unpaired electrons -- outer complex
(c) 5 d-electrons -- strong ligand -- 1 unpaired electron -- inner complex
(d) 6 d-electrons -- strong ligand -- no unpaired electrons -- inner complex
(e) 3 d-electrons -- weak ligand -- 3 unpaired electrons -- inner complex

21.95 $[Co(NO_2)_6]^{4-}$ should be easy to oxidize to $[Co(NO_2)_6]^{3-}$ because the cobalt in $[Co(NO_2)_6]^{4-}$ has an electron in the e_g level (high energy electron) while the cobalt in $[Co(NO_2)_6]^{3-}$ does not.

22 HYDROGEN, OXYGEN, NITROGEN, AND CARBON, AND AN INTRODUCTION TO ORGANIC CHEMISTRY

22.1 The official name for H_2 is dihydrogen.

22.3 Hydrogen is less abundant on Earth than the rest of the universe because the Earth's gravity wasn't strong enough to hold on to the hydrogen.

22.5 Advantage: less dense than He, so it has greater lifting power
Disadvantage: hydrogen is extremely flammable but helium doesn't burn.

22.7 ^1_1H, protium; ^2_1H or D, deuterium; ^3_1H or T, tritium. Tritium is radioactive and is dangerous because it can replace ordinary hydrogen in molecules that can be incorporated into the body.

22.9 (a) $CH_4 + H_2O \xrightarrow[\text{catalyst}]{\text{heat}} CO + 3H_2$ $\quad CO + H_2O \xrightarrow[\text{catalyst}]{\text{heat}} CO_2 + H_2$

(b) $C + H_2O \xrightarrow{1000°C} CO + H_2$

22.11 Hydrogen is a by-product in the preparation of caustic soda.

22.13 $Zn + H_2SO_4 \rightarrow ZnSO_4 + H_2$ See Figure 22.2.

22.15 $H{-}C{\equiv}C{-}H + 2H{-}H \longrightarrow CH_3{-}CH_3$

Hydrogenation of vegetable oils gives solid (or semi-solid) fats.

22.17 Hydrogen normally forms only one covalent bond. It has only one valence electron and needs only one more to fill its valence shell, therefore, it usually forms only the single bond.

22.19 As an acid-base reaction:

H^-	+	H_2O	→	H_2	+	OH^-
base		acid		acid		base

As a redox reaction:

H^- is oxidized to give H_2; H^+ in H_2O is reduced.

22.21 Some nonmetal hydrides have positive $\Delta G°_f$ and cannot be made by direct combination of elements. Those that can be made in high yield by direct combination are: CH_4, NH_3, H_2O, H_2S, HF, HCl, and HBr.

22.23 The atmosphere is the commercial source of oxygen and nitrogen.

22.25 In the laboratory oxygen is conveniently prepared by the catalytic decomposition of $KClO_3$.

22.27

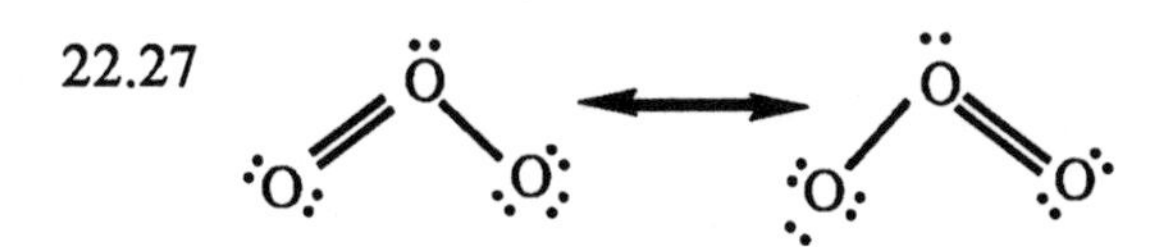

Ozone is made in the laboratory by electric discharge through O_2.
Ozone has the advantage over Cl_2 for purification of drinking water because O_3 doesn't form poisonous and carcinogenic compounds with impurities in the water, but it does kill bacteria.

22.29 $O^{2-} + H_2O \xrightarrow{100\%} 2OH^-$

22.31 $Al_2O_3 + 6H^+ \rightarrow 2Al^{3+} + 3H_2O$

$Al_2O_3 + 2OH^- \rightarrow 2AlO_2^- + H_2O$

22.33 The $\Delta G°_f$ for N_2O is positive; therefore, $K_c \ll 1$ for its formation and it cannot be prepared in high yields from its elements.

22.35 (a) $Cu + 4HNO_3 \rightarrow Cu(NO_3)_2 + 2NO_2 + 2H_2O$ (concentrated)

(b) $3Cu + 8HNO_3 \rightarrow 3Cu(NO_3)_2 + 2NO + 4H_2O$ (dilute)

22.37 $H-\ddot{\underset{\cdot\cdot}{O}}-\ddot{\underset{\cdot\cdot}{O}}-H$, see Figure 22.6.

22.39 The lack of reactivity of N_2 is attributed to the strength of the nitrogen-nitrogen triple bond.

22.41 Nitrogen-fixing bacteria remove N_2 from the air and make usable nitrogen compounds from it.

22.43 Lithium is the only element that reacts with N_2 at room temperature.

22.45 As a result of forming hydrogen bonds with water, ammonia is very soluble in water.

22.47 Ammonia can be prepared in the laboratory by the following reaction:
$2NH_4Cl + CaO \rightarrow 2NH_3 + H_2O + CaCl_2$
Ammonia can't be collected by displacement of water because it is too soluble in water.

22.49 $4NH_3(g) + 5O_2(g) \xrightarrow{Pt} 4NO(g) + 6H_2O(g)$

$2NO(g) + O_2(g) \rightarrow 2NO_2(g)$

$3NO_2(g) + H_2O(\ell) \rightarrow 2HNO_3(\ell) + NO(g)$

22.51 Nitric acid can be prepared in the laboratory by the following reaction:
$NaNO_3 + H_2SO_4 + heat \rightarrow NaHSO_4 + HNO_3$

22.53 When zinc reacts with nitric acid, NH_4^+ is formed.

22.55 Aqua regia is one part HNO_3 and three parts HCl, by volume. Nitrate ion oxidizes the noble metal, and chloride ion acts as a complex ion-forming agent to shift the equilibrium toward the dissolved metal.

22.57 $N_2 + 3H_2 \xrightarrow[\text{catalyst}]{\text{iron}} 2NH_3$

Temp; 400-500°C. Pressure; several hundred atmospheres. These conditions were chosen to produce the maximum amount of NH_3 in the shortest time. Pressure shifts the equilibrium to more product and temperature increases the reaction rate.

22.59 A disproportionation reaction is one in which the same chemical undergoes both oxidation and reduction.

22.61 $\text{H}-\ddot{\text{N}}(\text{H})-\ddot{\text{N}}(\text{H})-\text{H}$, see Figure 22.9

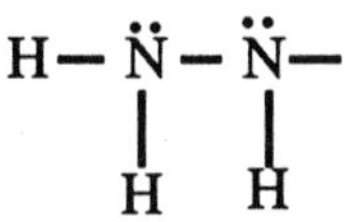

22.63 Hydroxylamine is: $\text{H}-\ddot{\text{N}}(\text{H})-\ddot{\underset{\cdot\cdot}{\text{O}}}-\text{H}$ Basicities increase: $NH_2OH < N_2H_4 < NH_3$.

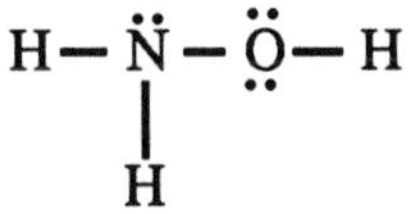

22.65 The preparation of nitrous oxide involves:

$$NH_4NO_3(\ell) \xrightarrow{\text{heat}} N_2O(g) + 2H_2O(g)$$

The resonance structures for nitrous oxide are:

$$:\ddot{\text{N}}{=}\text{N}{=}\ddot{\text{O}}: \longleftrightarrow :\text{N}{\equiv}\text{N}{-}\ddot{\underset{\cdot\cdot}{\text{O}}}:$$

22.67 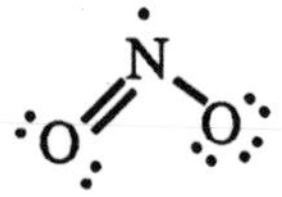 and

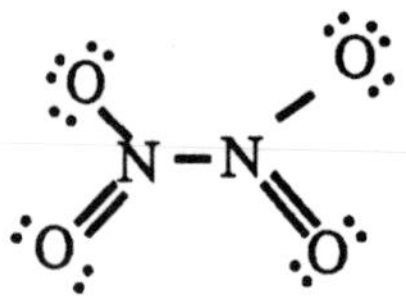

$2NO_2 \rightleftharpoons N_2O_4$

22.69 $3HNO_2 \rightarrow HNO_3 + H_2O + 2NO$

22.71 (a) As a reducing agent: $H^+ + 5HNO_2 + 2MnO_4^- \rightarrow 5NO_3^- + 2Mn^{2+} + 3H_2O$

(b) As an oxidizing agent: $2H^+ + 2HNO_2 + 2I^- \rightarrow I_2 + 2NO + 2H_2O$

22.73 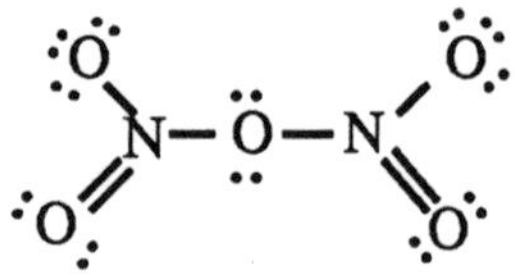 (vapor); $NO_2^+ NO_3^-$ in solid

22.75 $N_2O_5 + H_2O \rightarrow 2HNO_3$

22.77 A series of reactions that causes photochemical smog is:

$N_2 + O_2 \rightleftharpoons 2NO$

$2NO + O_2 \rightarrow 2NO_2$

$NO_2 \xrightarrow{hv} NO + O$

$O + O_2 \rightarrow O_3$

O_3 + hydrocarbons $\rightarrow$ PAN

22.79 Heating coal in the absence of air produces coke.

22.81 Graphite is used as a lubricant and in electrodes.

22.83 Carbon formed by heating wood in the absence of air and then being finely pulverized is known as activated charcoal. Its large surface area per unit mass allows it to adsorb large numbers of molecules.

22.85 $:C{\equiv}O:$, $\ddot{\underset{..}{O}}{=}C{=}\ddot{\underset{..}{O}}$

22.87 $Fe_2O_3(s) + 3CO(g) \xrightarrow{heat} 2Fe(s) + 3CO_2(g)$

22.89 Commercially CO_2 is made from limestone.

$$CaCO_3 \xrightarrow{heat} CaO + CO_2$$

In the laboratory it is made by the reaction:

$$CaCO_3(s) + 2H^+(aq) \rightarrow Ca^{2+}(aq) + CO_2(g) + H_2O(\ell)$$

22.91 Green plants use CO_2 during photosynthesis to make glucose from which they make cellulose, starch and other chemicals.

22.93 Increased CO_2 in the Earth's atmosphere will absorb greater amounts of infrared radiation that would otherwise be radiated into space. The absorbed radiation will be converted to heat and cause a gradual warming known as the greenhouse effect.

22.95 As H_2O evaporates, the following equilibrium is shifted to the right:

$$Ca(HCO_3)_2(aq) \rightleftharpoons CaCO_3(s) + H_2O(\ell) + CO_2(g)$$

and solid $CaCO_3$ forms the stalagmites and stalactites.

22.97 Boiler scale can be removed by washing with dilute acid to dissolve the $CaCO_3$.

22.99 $Al_4C_3(s) + 12H_2O(\ell) \rightarrow 4Al(OH)_3(s) + 3CH_4(g)$

22.101 An interstitial carbide contains carbon atoms which are located between atoms of the host lattice. An example is tungsten carbide, WC.

22.103 $\ddot{\underset{..}{S}}=C=\ddot{\underset{..}{S}}$; CS_2 is very flammable.

22.105 Unsaturated hydrocarbons contain one or more double and/or triple bonds between the carbons. Saturated compounds contain only single bonds between carbons.

22.107 (a) $C_{30}H_{62}$ (b) $C_{27}H_{54}$ (c) $C_{33}H_{64}$

22.109 This compound is one of the 13 isomers of hexene. These are shown in the answer to Question 22.113.

22.111

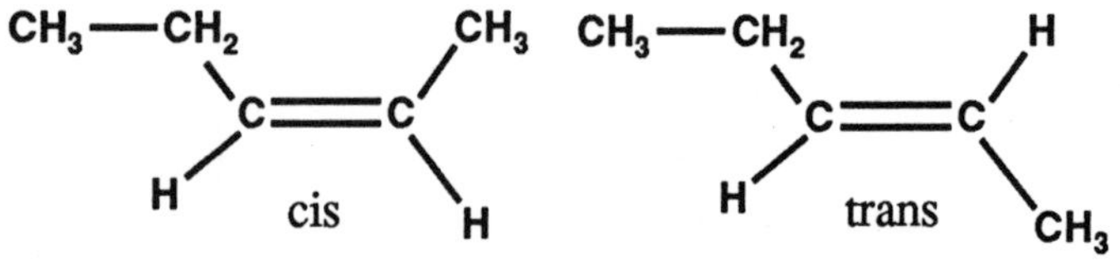

```
CH3—CH2          CH3      CH3—CH2          H
       \        /                \        /
        C======C                  C======C
       /        \                /        \
      H   cis    H              H  trans   CH3
```

22.113 There are 13 isomers:

```
       H    H    H    H    H
       |    |    |    |    |      H
  H —  C —  C —  C —  C —  C = C/
       |    |    |    |          \H
       H    H    H    H
```

1-hexene

```
       H    H    H    H    H    H
       |    |    |    |    |    |
  H —  C —  C —  C —  C =  C —  C — H
       |    |    |              |
       H    H    H              H
```

2-hexene; gives geometrical isomers

```
       H    H    H    H    H    H
       |    |    |    |    |    |
  H —  C —  C —  C =  C —  C —  C — H
       |    |              |    |
       H    H              H    H
```

3-hexene; gives geometrical isomers

```
       H    H    H
       |    |    |              H
  H —  C —  C —  C —  C =  C/
       |    |    |    |      \H
       H    H    H    C
                    H ' H
                      H
```

2-methyl-1-pentene

```
       H    H    H    H
       |    |    |    |         H
  H —  C —  C —  C* — C =  C/
       |    |    |           \H
       H    H    C
               H ' H
                 H
```

3-methyl-1-pentene; gives optical isomers

```
       H    H    H    H
       |    |    |    |         H
  H —  C —  C —  C —  C =  C/
       |    |    |           \H
       H    C    H
          H ' H
            H
```

4-methyl-1-pentene

(continued)

22.113 (continued)

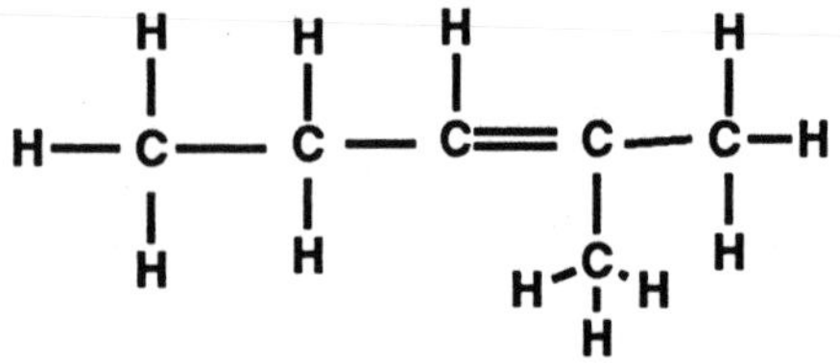

2-methyl-2-pentene

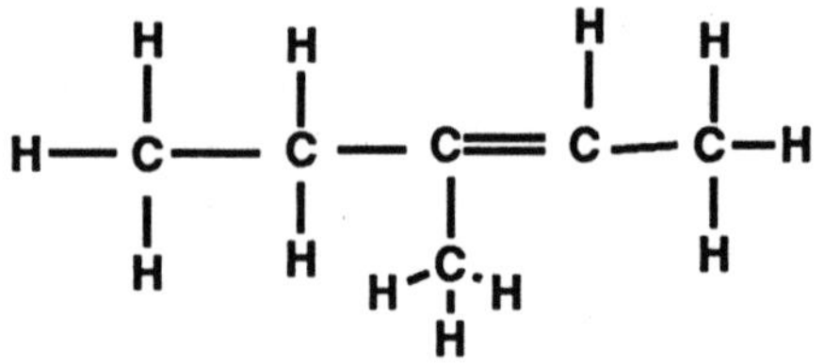

3-methyl-2-pentene; gives geometrical isomers

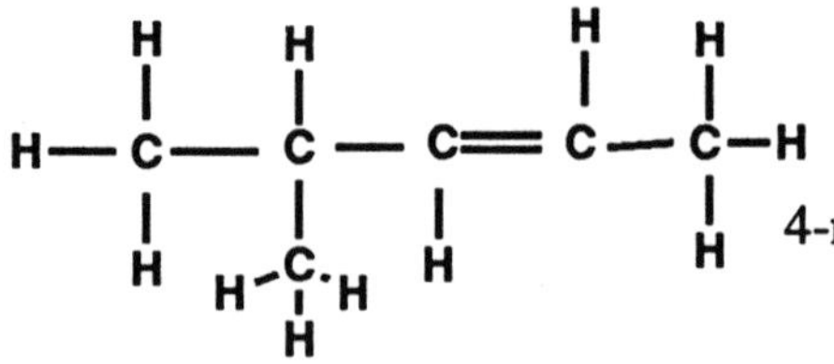

4-methyl-2-pentene; gives geometrical isomers

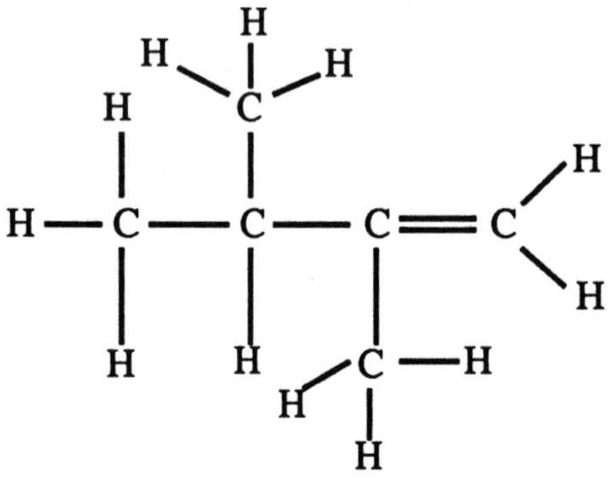

2,3-dimethyl-1-butene

(continued)

22.113 (continued)

```
             H
         H   |   H
    H     \  |  /
    |      \ C /                 H
    |        |                  /
H — C ————— C ——— C ═══ C             3,3-dimethyl-1-butene
    |        |     \      \
    |        |      \      H
    H    H — C — H   H
             |
             H
```

```
    H       H
    |       |                   H
    |       |                  /
H — C ————— C ——— C ═══ C             2-ethyl-1-butene
    |       |     |      \
    |       |     |       H
    H       H H — C — H
                  |
              H — C — H
                  |
                  H
```

```
    H                           H
    |                           |
    |                           |
H — C ——— C ═══ C ——— C — H          2,3-dimethyl-2-butene
    |     |     |     \
    |     |     |      H
    H     | H — C — H
      H — C — H |
          \     |
           H    H
```

22.115

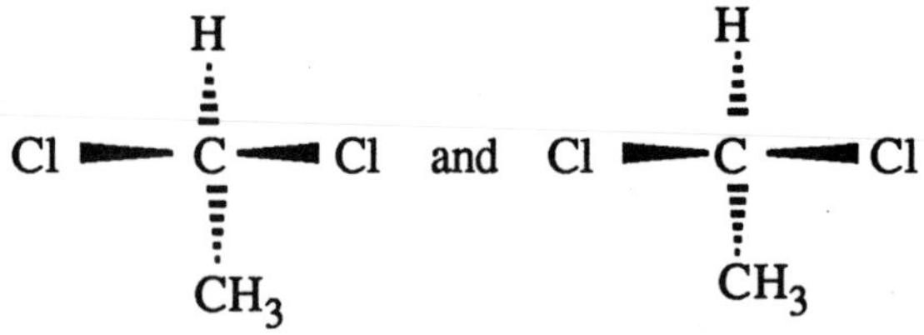

$$\mathrm{Cl} \blacktriangleright \underset{\mathrm{CH_3}}{\overset{\mathrm{H}}{\mathrm{C}}} \blacktriangleleft \mathrm{Cl} \quad \text{and} \quad \mathrm{Cl} \blacktriangleright \underset{\mathrm{CH_3}}{\overset{\mathrm{H}}{\mathrm{C}}} \blacktriangleleft \mathrm{Cl}$$

These are mirror images that are superimposable; therefore, they are identical and not chiral.

22.117 C_2Cl_4 is planar because of the double bond between the carbons. Each carbon uses sp^2 hybrid orbitals which gives a planar configuration and there is no rotation about the double bond.

22.119 (a) 2,4-dimethylhexane

(b) 3,5-dimethylheptane

(c) 5-ethyl-3-methyloctane

(d) 5-methyl-3-heptene

(e) 2,4-dimethylhexane

22.121 (a) 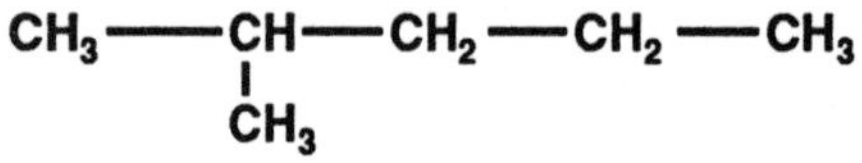 $CH_3—\underset{\substack{|\\CH_3}}{CH}—CH_2—CH_2—CH_3$

(b) 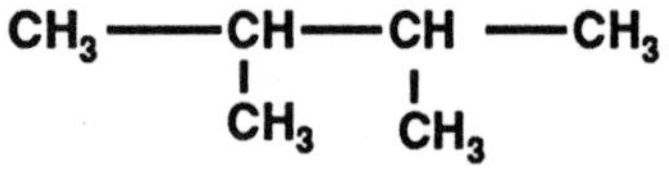 $CH_3—\underset{\substack{|\\CH_3}}{CH}—\underset{\substack{|\\CH_3}}{CH}—CH_3$

(c) 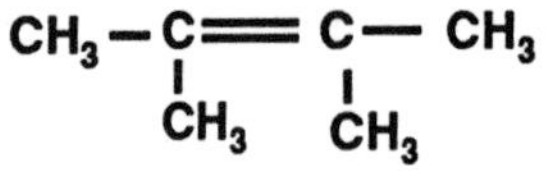 $CH_3—\underset{\substack{|\\CH_3}}{C}=\underset{\substack{|\\CH_3}}{C}—CH_3$

(continued)

22.121 (continued)

(d) $CH_2{=}CH{-}CH{=}CH{-}CH{=}CH{-}CH_2{-}CH_3$

(e) 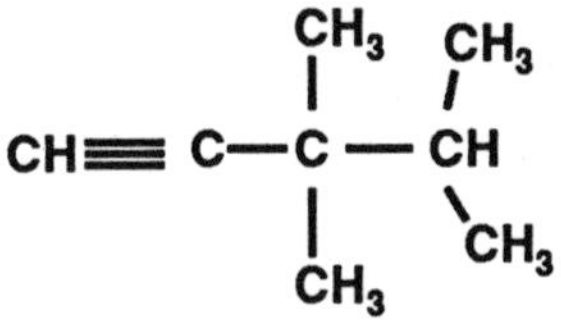 $CH{\equiv}C{-}C(CH_3)_2{-}CH(CH_3)_2$

22.123 A functional group is an atom or group of atoms that bestows some characteristic property to a molecule so that any molecule with the same grouping will react chemically in a similar fashion.

(a) 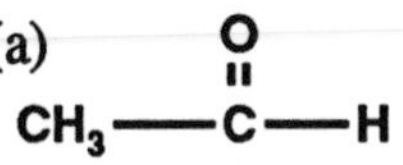 $CH_3{-}C({=}O){-}H$ (b) $CH_3{-}C({=}O){-}CH_3$ (c) $CH_3{-}C({=}O){-}OH$ (d) $CH_3{-}NH_2$

(e) $CH_3{-}CH_2{-}OH$ (f) 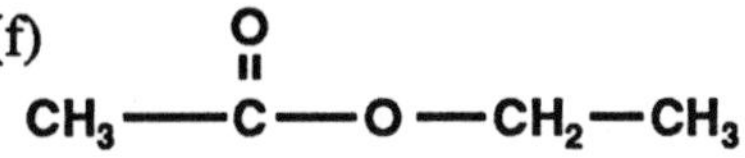 $CH_3{-}C({=}O){-}O{-}CH_2{-}CH_3$ (g) $CH_3{-}O{-}CH_3$

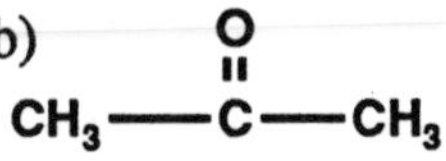

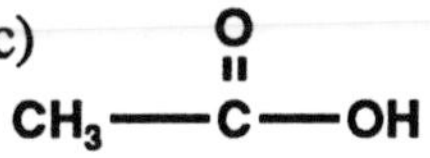

22.125 (a) 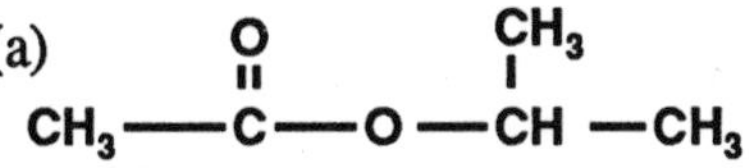 $CH_3{-}C({=}O){-}O{-}CH(CH_3){-}CH_3$ (b) $CH_3{-}C({=}O){-}O{-}CH_2CH_2CH_2CH_2CH_3$

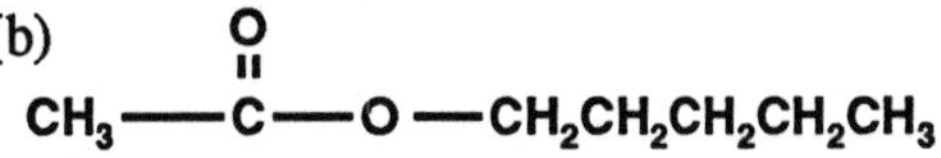

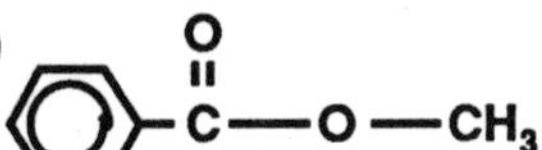

(c) $C_6H_5{-}C({=}O){-}O{-}CH_3$ (d) 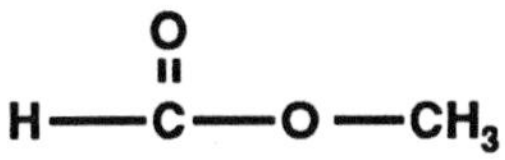 $H{-}C({=}O){-}O{-}CH_3$

22.127 Esters

22.129 Alkenes and alkynes tend to undergo addition reactions whereas alkanes tend to undergo substitution reactions.

22.131 (a)

$$CH_3CH_2-\overset{O}{\overset{\|}{C}}-OCH_3 + NaOH \xrightarrow{H_2O} CH_3OH + NaO-\overset{O}{\overset{\|}{C}}-CH_2CH_3$$

(b)

$$CH_3CH_2O\text{-}\overset{O}{\overset{\|}{C}}CH_2CH_2\overset{O}{\overset{\|}{C}}\text{-}OCH_2CH_3 + 2NaOH$$

$$\xrightarrow{H_2O} 2CH_3CH_2OH + NaO\overset{O}{\overset{\|}{C}}CH_2CH_2\overset{O}{\overset{\|}{C}}ONa$$

23 PHOSPHORUS, SULFUR, THE HALOGENS, THE NOBLE GASES, AND SILICON, AND AN INTRODUCTION TO POLYMER CHEMISTRY

23.1 $Ca_3(PO_4)_2$

23.3 $[Ne]3s^23p^3$

23.5 $2Ca_3(PO_4)_2(s) + 6SiO_2(s) + 10C(s) \xrightarrow{1300^\circ C} 6CaSiO_3(\ell) + 10CO(g) + P_4(g)$

23.7 **Red phosphorus** - believed to consist of P_4 tetrahedra joined to each other at their corners. **Black phosphorus** - layers of phosphorus atoms in which atoms in a given layer are covalently bonded to each other. Binding between layers is weak. Both of these forms are less reactive than white phosphorus.

23.9 (a) See Figure 23.2 (b) See Figure 23.3

23.11 A desiccant is a substance that removes H_2O from a gas mixture. P_4O_{10} reacts with water to give H_3PO_4, thus giving a moisture-free gas.

23.13 Concentrated H_3PO_4 that is about 85% H_3PO_4 by weight is called syrupy phosphoric acid.

23.15 The manufacture of fertilizers, food additives, and detergents have been three very large uses of phosphoric acid.

23.17 H_3PO_4

23.19 TSP is used as a water softener and as a cleansing agent. Solutions of Na_3PO_4 are basic because the hydrolysis of PO_4^{3-} makes the solution basic. $PO_4^{3-} + H_2O \rightarrow HPO_4^{2-} + OH^-$

23.21 Superphosphate fertilizer is a mixture of calcium sulfate and calcium dihydrogen phosphate. It is made by the following reaction:

$$Ca_3(PO_4)_2 + 2H_2SO_4 + 4H_2O \rightarrow 2CaSO_4{\bullet}2H_2O + Ca(H_2PO_4)_2$$

23.23 (a) sodium hydrogen phosphate (b) sodium dihydrogen phosphate

23.25 See the Lewis structures in the portion of Section 23.1 on "Polymeric Phosphoric Acids and Their Anions" that illustrates the formation of pyrophosphoric acid.

23.27 $(PO_3^-)_n$ is the metaphosphate ion. Its empirical formula is PO_3^-. It is formed by condensation of $H_2PO_4^-$. See Section 23.1 for an illustration of how it can be considered to be formed from phosphoric acid. For its Lewis structure the non-bonding electrons need to be added to an illustration like that referred to in Section 23.1.

23.29 3 mol NaH_2PO_4 to 2 mol Na_2HPO_4 (the HPO_4^{2-} units terminate the ends of the chains).

23.31 Phosphorous acid. $Mg(H_2PO_3)_2$ and $MgHPO_3$ (H_3PO_3 is a diprotic acid).

23.33 See Figure 23.4. PCl_5 exists as $PCl_4^+PCl_6^-$ in the solid. In PCl_3, phosphorus uses sp^3 hybrid orbitals; in PCl_5 vapor or liquid it uses sp^3d hybrid orbitals. In solid PCl_4^+ it uses sp^3 and PCl_6^- it uses sp^3d^2.

23.35 Brimstone is another name for elemental sulfur. It means "stone that burns."

23.37 The two allotropic forms of sulfur are rhombic sulfur and monoclinic sulfur. They have different packing of S_8 rings in their crystals. When heated, solid sulfur melts to give an amber liquid containing S_8 rings. The rings break and join with S_x chains as the liquid darkens and thickens. The S_x chains break into smaller fragments at still higher temperatures and becomes less viscous again.

23.39 Superheated water is pumped into the sulfur deposit where it melts the sulfur. This is then foamed to the surface with compressed air. See Figure 23.5.

23.41 $S(s) + O_2(g) \rightarrow SO_2(g)$ $\qquad$ $2SO_2(g) + O_2(g) \xrightarrow{\text{catalyst}} 2SO_3(g)$

$SO_3(g) + H_2SO_4(\ell) \rightarrow H_2S_2O_7(\ell)$ $\qquad$ $H_2S_2O_7(\ell) + H_2O(\ell) \rightarrow 2H_2SO_4(\ell)$

23.43 SO_2 can be conveniently prepared in the laboratory by reaction of a sulfite (e.g., Na_2SO_3) with an acid.

$Na_2SO_3(s) + H_2SO_4(aq) \rightarrow Na_2SO_4(aq) + H_2O(\ell) + SO_2(g)$

23.45 Rain falling through air that is polluted with SO_2 and SO_3 from the burning of sulfur-containing fuels becomes acidic because SO_2 and SO_3 react with water to form H_2SO_3 and H_2SO_4. It causes structural damage to buildings, causes corrosion of metals, and kills fish and plants.

23.47 Sulfuric acid is used in the production of fertilizers, in refining petroleum, in lead storage batteries, in the manufacture of other chemicals, and by the steel industry.

23.49 Add concentrated H_2SO_4 to the water -- never the other way around.

23.51 $C_6H_{12}O_6 \xrightarrow[\text{conc.}]{H_2SO_4} 6C + 6H_2O$

23.53 Look at the structure of pyrophosphoric acid then give the Lewis structure of the formula, $H[OSO_2]_2OH$

23.55 $CH_3CSNH_2 + 2H_2O \rightarrow CH_3CO_2^- + NH_4^+ + H_2S$

23.57 The structure of the thiosulfate ion is given near the end of Section 23.2.
$S(s) + SO_3^{2-}(aq) \rightarrow S_2O_3^{2-}(aq)$

23.59 (a) $S_2O_3^{2-} + 4Cl_2 + 5H_2O \rightarrow 8Cl^- + 2SO_4^{2-} + 10H^+$
(b) $2S_2O_3^{2-} + I_2 \rightarrow S_4O_6^{2-} + 2I^-$

23.61 F, $1s^22s^22p^5$; Cl, $1s^22s^22p^63s^23p^5$; Br, $1s^22s^22p^63s^23p^63d^{10}4s^24p^5$;
I, $1s^22s^22p^63s^23p^63d^{10}4s^24p^64d^{10}5s^25p^5$

23.63 Halogens are normally found in nature in the combined state, normally as halide ions.

23.65 Fluorine is prepared by the electrolysis of HF dissolved in molten KF.

23.67 F_2: Pale yellow gas, b.p. of -188°C m.p. -233°C
Cl_2: Pale yellow-green gas, b.p. = -34.6°C m.p. -103°C
Br_2: Dark red liquid, b.p. = 58.8°C m.p. -7.2°C
I_2: Dark, metallic-looking solid, b.p. 184.4°C m.p. = 113.5°C

23.69 Elemental fluorine is very reactive as an oxidizing agent because the F–F bond splits easily to form 2 very reactive F atoms.

23.71 Iodine is recovered from seaweed and from $NaIO_3$, which is an impurity in Chilean saltpeter.

23.73 $CaF_2(s) + H_2SO_4(\ell) \rightarrow CaSO_4(s) + 2HF(g)$

23.75 $NaBr + H_3PO_4 \xrightarrow{heat} HBr + NaH_2PO_4$

$NaI + H_3PO_4 \xrightarrow{heat} HI + NaH_2PO_4$

23.77 [Boiling Points] $HCl < HBr < HI < HF$
In the liquid state HF is hydrogen bonded into staggered chains.

23.79 (a) hypobromous acid (b) sodium hypochlorite (c) potassium bromate
(d) magnesium perchlorate (e) periodic acid (f) bromic acid
(g) sodium iodate (h) potassium chlorite

23.81 (a) $Cl_2 + H_2O \rightleftharpoons H^+ + Cl^- + HOCl$ (about 30% of chlorine as HOCl and Cl^-)

(b) $Cl_2 + 2OH^- \rightarrow OCl^- + Cl^- + H_2O$ (equilibrium lies further to the right)

23.83 OCl^- is stable, OBr^- reacts moderately fast, OI^- reacts very rapidly. Stability of OCl^- is related to slow rate of reaction rather than to thermodynamic stability.

23.85 $Ca(OCl)_2$

23.87 Seven fluorine atoms cannot fit around the smaller chlorine atom.

23.89 $GeCl_4 + 2H_2O \rightarrow GeO_2 + 4HCl$ (like the reaction of $SiCl_4$)

23.91 NH_3 is a good Lewis base; NF_3 is a poor Lewis base because the highly electronegative fluorine atoms make the nitrogen in NF_3 a poor electron pair donor.

23.93 XeF_4, square planar with two nonbonding pairs of electrons; XeF_2, linear with three nonbonding pairs of electrons;.

23.95 $SiO_2(s) + 2C(s) \xrightarrow{\text{heat}} Si(s) + 2CO(g)$

23.97 Si does not form stable π-bonds to other Si atoms.

23.99 The structure of orthosilicate ion is given in Section 23.5.

23.101 Draw a Lewis structure like that in the section on "Compounds With Silicon Oxygen Bonds" with sufficient repeating SiO_3^{2-} units to give a total of six Si atoms (See Figure 23.16 b). It is found in beryl, $Be_3Al_2(Si_6O_{18})$ and emeralds.

23.103 See Figure 23.17

23.105 SiO_2 is the empirical formula for quartz.

23.107 The chemical properties of polymers are similar to the chemical properties of the functional groups of the smaller, repeating units when they were not contained in such a large molecule. The physical properties of polymers are reflections of their enormous size.

23.109 Addition polymerization reactions involve the opening of double bonds and the pairing of electrons between the monomer units to create the links. An initiator is used to get the process of polymerization started.

23.111 Teflon is an especially useful polymer because it is very resistant to chemical attack, very slippery and easy to cleanup (non-stick) because substances do not adhere to it.

23.113 The bond between nitrogen and carbon in the NH-CO group (from an amine and a carboxylic acid) is the amide bond. The amide bond is found in peptides (proteins) as well as in polymers. The amide bond is present in nylon.

23.115 Repeat a portion of the drawing found in the section on "Phosphazenes." If the groups attached to the backbone of a polyphosphazene polymer are polar, a water-soluble polymer is obtained; if they are nonpolar, the polymer is water-insoluble.

23.117 A portion of the structure of a silicone polymer is shown in the section on "Silicone Polymers."

23.119 Simethicone is a medicinally used polymer and is shown as the product of the last equation in the section on "Silicone Polymers."

23.121 There is interest in polymers like poly(sulfur nitride) because of their ability to conduct electricity and to act as superconductors.

23.123 Polysaccharides are polymers of sugar molecules.

23.125 Nucleotides contain the three units, deoxyribose (a sugar), a nitrogenous base, and the phosphoric acid unit. They are linked as shown in Figure 23.29 with the phosphate unit as the bridge between the sugar units.

24 NUCLEAR CHEMISTRY

24.1 Alpha particles, beta particles and gamma rays are the three main types of radiation emitted by radioactive nuclei. See Table 24.1

24.3 (a) $^{81}_{36}Kr + ^{0}_{-1}e \rightarrow ^{81}_{35}Br$

(b) $^{104}_{47}Ag \rightarrow ^{0}_{1}e + ^{104}_{46}Pd$

(c) $^{73}_{31}Ga \rightarrow ^{0}_{-1}e + ^{73}_{32}Ge$

(d) $^{104}_{48}Cd \rightarrow ^{104}_{47}Ag + ^{0}_{1}e$

(e) $^{54}_{25}Mn + ^{0}_{-1}e \rightarrow ^{54}_{24}Cr$

24.5 (a) $^{135}_{53}I \rightarrow ^{135}_{54}Xe + ^{0}_{-1}e$

(b) $^{245}_{97}Bk \rightarrow ^{4}_{2}He + ^{241}_{95}Am$

(c) $^{238}_{92}U + ^{12}_{6}C \rightarrow ^{246}_{98}Cf + 4\,^{1}_{0}n$

(d) $^{96}_{42}Mo + ^{2}_{1}H \rightarrow ^{1}_{0}n + ^{97}_{43}Tc$

(e) $^{20}_{8}O \rightarrow ^{20}_{9}F + ^{0}_{-1}e$

24.7 (a) ${}^{11}_{5}B \rightarrow {}^{4}_{2}He + {}^{7}_{3}Li$

(b) ${}^{98}_{38}Sr \rightarrow {}^{0}_{-1}e + {}^{98}_{39}Y$

(c) ${}^{107}_{47}Ag + {}^{1}_{0}n \rightarrow {}^{108}_{47}Ag$

(d) ${}^{88}_{35}Br \rightarrow {}^{1}_{0}n + {}^{87}_{35}Br$

(e) ${}^{116}_{51}Sb + {}^{0}_{-1}e \rightarrow {}^{116}_{50}Sn$

(f) ${}^{70}_{33}As \rightarrow {}^{0}_{1}e + {}^{70}_{32}Ge$

(g) ${}^{41}_{19}K \rightarrow {}^{1}_{1}H + {}^{40}_{18}Ar$

24.9 In a radioactive decay, the isotope that decays is called the parent isotope and the isotope that is formed is referred to as the daughter isotope.

24.11 $\ln \frac{[A]_0}{[A]_t} = kt$ (Equation 24.1)

$[A]_t = 1/2[A]_0$ @ $t_{1/2}$

$$\ln \frac{[A]_0}{1/2[A]_0} = kt_{1/2}$$

$\ln(2) = kt_{1/2}$ or $t_{1/2} = \frac{\ln(2)}{k}$ $\quad$ $\mathbf{t_{1/2} = \frac{0.693}{k}}$ (Equation 24.2)

24.13 (See Problem 24.12)

(a) $\ln \frac{8.00\ g}{X_a} = \frac{240}{120}$ (half-life periods) x 0.693 = 1.386

$\mathbf{X_a = 2.00\ g}$

(b) (4 half-life periods) $\mathbf{X_b = 0.500\ g}$ $\quad$ (c) (8 half-lives) $\mathbf{X_c = 0.0312\ g}$

24.15 $t_{1/2} = \frac{0.693}{k} = \frac{0.693}{2.30 \times 10^{-6}\ \text{year}^{-1}} = \mathbf{3.01 \times 10^{5}\ years}$

24.17 $k = \frac{0.693}{t_{1/2}} = \frac{0.693}{470\ \text{days}} = \mathbf{0.00147\ day^{-1}}$

24.19 See Figure 24.4 and the first part of the section on "Measurement of Radioactivity".

24.21 The rem (radiation equivalent for man) takes into consideration the amount of damage the various forms of radiation can cause to animal tissue. Rad does not differentiate between types of radiation.

24.23 (a) $\frac{140\ \text{Bq}}{\text{g}} \times 1.0\ \text{mg} \times \frac{1.00\ \text{g}}{1{,}000\ \text{mg}} \times \frac{1.00\ \text{dis s}^{-1}}{\text{Bq}} \times 10\ \text{min}$

$$\times \frac{60\ \text{s}}{\text{min}} \times \frac{1.80\ \text{MeV}}{\text{dis}} \times \frac{1.60 \times 10^{-13}\ \text{J}}{\text{MeV}} \times \frac{1}{\text{body weight}}$$

Answer = $\mathbf{2.4 \times 10^{-11}}$ **J/body weight** If we assume the average body weight is 60 kg, then the answer is 4.0×10^{-13} J/kg = $\mathbf{4 \times 10^{-13}}$ **Gy**

(b) $\frac{(4.0 \times 10^{-13}\ \text{Gy})}{(\text{kg of body weight})} \times \frac{100\ \text{rad}}{\text{Gy}} = \mathbf{4.0 \times 10^{-11}\ rad}$ (for a 60 kg weight person)

24.25 mol ^{40}Ar formed = mol ^{40}K decayed = 1.15×10^{-5} mol

$k = \frac{0.693}{t_{1/2}} = \frac{0.693}{1.3 \times 10^{9}\ \text{yr}} = 5.3 \times 10^{-10}\ \text{yr}^{-1}$ $t = \frac{2.303}{k} \times \log\frac{[A]_0}{[A]_t}$

$[A]_0 = (2.07 \times 10^{-5} + 1.15 \times 10^{-5})$ mol ^{40}K $[A]_t = 2.07 \times 10^{-5}$ mol ^{40}K

(continued)

24.25 (continued)

$$t = \frac{2.303}{5.3 \times 10^{-10}\ yr^{-1}} \times \log \frac{3.22 \times 10^{-5}}{2.07 \times 10^{-5}} = \mathbf{8.3 \times 10^{8}\ yr}$$

24.27 ON + O*NO ⇌ ONO*NO ⇌ ONO + *NO

bond breaking at this point to yield these products

bond breaking at this point to yield these products

24.29 One possible experiment would be to make the complex and allow the racemization to occur in a medium containing labeled $C_2O_4^{2-}$. If the racemization occurs by the loss and recombination with $C_2O_4^{2-}$, the complex should pick up labeled $C_2O_4^{2-}$ during the racemization.

24.31 $?\text{mol Cr} = 165\ \text{cpm} \times \frac{1\ g\ K_2Cr_2O_7}{843\ \text{cpm}} \times \frac{2\ \text{mol Cr}}{294\ g\ K_2Cr_2O_7} = 1.33 \times 10^{-3}\ \text{mol Cr}$

$$?\text{mol } C_2O_4^{2-} = 83\ \text{cpm} \times \frac{1\ g\ H_2C_2O_4}{345\ \text{cpm}} \times \frac{1\ \text{mol } C_2O_4^{2-}}{90.0\ g\ H_2C_2O_4} = 2.67 \times 10^{-3}\ \text{mol } C_2O_4^{2-}$$

Therefore, there are two oxalate ions bound to each Cr(III) in the complex ion.

24.33 Elements higher than 83 must lose both neutrons and protons to achieve a stable n/p ratio. The only way this is possible is by α-emission or fission.

24.35 e = even, o = odd, * = magic number

$${}^{4}_{2}He > {}^{58}_{28}Ni > {}^{39}_{20}Ca > {}^{71}_{32}Ge > {}^{10}_{5}B$$

(p,n) (e*,e*) (e*,e) (e*,o) (e,o) (o,o)

24.37 Radiation emitted in quantum packets can be used to explain nuclear shells just as Bohr did in explaining his atomic theory. If protons move from shell to shell in the nucleus, each transition would result in an emission of a different energy.

24.39 A nuclear transformation is a nuclear reaction in which a bombarding particle is absorbed and causes the absorbing nucleus to change into a nucleus of another element.

24.41 Transuranium elements are elements 93 to 105 (elements with atomic numbers greater than that of uranium). Transuranium elements do not occur in nature, they have only been produced artificially.

24.43 (a) ${}^{242}_{96}Cm + {}^{4}_{2}He \rightarrow {}^{245}_{98}Cf + {}^{1}_{0}n$ (b) ${}^{108}_{48}Cd + {}^{1}_{0}n \rightarrow {}^{109}_{48}Cd + \gamma$

(c) ${}^{14}_{7}N + {}^{1}_{0}n \rightarrow {}^{14}_{6}C + {}^{1}_{1}H$ (d) ${}^{27}_{13}Al + {}^{2}_{1}H \rightarrow {}^{25}_{12}Mg + {}^{4}_{2}He$

(e) ${}^{249}_{98}Cf + {}^{18}_{8}O \rightarrow {}^{263}_{106}Unh + 4{}^{1}_{0}n$

24.45 (a) Na_2Uuh (b) H_2Uuh (c) $UuhO_2$ Since Po is a metalloid, element 116 (directly below it on the periodic table) would probably have metallic properties.

24.47 The bombarding nuclei have to contain a very large n/p ratio to place the products on the island of stability (Figure 24.9) no. p $\approx$ 16 and no. n $\approx$ 50 - 62. Light nuclei, however, contain n/p ratios of nearly 1.

24.49 (a) 148 (b) 125 (c) 103 [This is lawrencium (Lr)]

24.51 $^{56}Fe = 26\ p \times 1.007277\ u/p + 30\ n \times 1.008665\ u/n + 26\ e \times 5.4859 \times 10^{-4}\ u/e$
$= 56.463415$ u.
Since the actual atomic mass = 55.9349 u, the mass defect = 0.5285 u. What is the binding energy per nucleon?

(continued)

24.51 (continued)

$$?\text{MeV} = \frac{0.5285\text{ u}}{56\text{ nucleon}} \times \frac{931\text{ MeV}}{1\text{ u}} = \mathbf{\frac{8.79\ MeV}{nucleon}}$$

Since this is the <u>largest</u> value of binding energy per nucleon (the highest point on the curve in Figure 24.10), neither fission nor fusion of iron 56 can yield energy, i.e., produce products with a greater binding energy per nucleon.

24.53 $E = mc^2$ $E = 2 \times 9.1096 \times 10^{-31}$ kg $(2.9979 \times 10^8 \text{ m s}^{-1})^2 = \mathbf{1.6374 \times 10^{-13}\ J}$ **(1 kg m^2 s^{-2} = 1 J)**

24.55 Fission is the splitting of an atom into approximately equal parts. The fission of ^{235}U releases energy and neutrons that can cause other atoms of ^{235}U to undergo fission. The result could be a nuclear explosion.

24.57 A breeder reactor is one that is designed to produce more nuclear fuel than it consumes. The nuclear equation for the reaction that generates the nuclear fuel that is produced is:

$$^{238}_{92}U + ^{1}_{0}n \rightarrow ^{239}_{94}Pu + 2\,^{0}_{-1}e$$

24.59 Critical mass is the minimum amount of the fissile isotope required to sustain a chain reaction.

24.61 The only naturally-occurring fissile isotope is U-235.

24.63 It is difficult to build fusion reactors because fusion reactions have very high energies of activation and a very high temperature is required. A fusion reactor is difficult to construct since there is not an acceptable means of achieving the very high temperatures and containing this reaction.

24.65 Fusion reactions might be better sources of energy in the future since (1) they produce much more energy than do fission reactions, (2) they produce products that usually are not radioactive while most fission reactions produce radioactive products, and (3) the supply of fuel is virtually inexhaustible.

24.67 $C_8H_{18}(\ell) + 25/2O_2(g) \rightarrow 8CO_2(g) + 9H_2O(\ell)$

$8\ \Delta H_f(CO_2(g)) + 9\ \Delta H_f(H_2O(\ell)) - \Delta H_f(C_8H_{18}(\ell)) = \Delta H_{comb.}(C_8H_{18}(\ell))$

$8(-394\ kJ/mol) + 9(-286\ kJ/mol) - (-208.4\ kJ/mol) = -5520\ kJ$

$$(C_8H_{18}(\ell)) = 1\ mol\ ^4He \times \frac{2.3009 \times 10^9\ kJ}{1\ mol\ ^4He} \times \frac{1\ mol\ C_8H_{18}}{5520\ kJ}$$

$$\times \frac{114\ g\ C_8H_{18}}{1\ mol\ C_8H_{18}} \times \frac{1\ L\ C_8H_{18}}{703\ g\ C_8H_{18}} \times \frac{1\ gal\ C_8H_{18}}{3.79\ L\ C_8H_{18}}$$

$$= \mathbf{1.78 \times 10^4\ gal\ C_8H_{18}(\ell)}$$

APPENDIX B:

MOLECULAR ORBITAL THEORY

B.1 From the point of view of molecular orbital theory, the electronic structure of a molecule is similar to the electronic structure of an atom in that (1) no more than two electrons can populate a single orbital, (2) each electron will occupy the lowest energy orbital available, and (3) electrons are spread out as much as possible over orbitals of the same energy with unpaired spins if possible. The electronic structures of atoms and molecules differ in that (1) atomic orbitals and molecular orbitals are very different and (2) in molecular orbital theory, distinctly different bonding and antibonding orbitals exist unlike anything that exists in the electronic structure of an atom.

B.3 See Figure B. 2.

B.5 The molecular orbital diagram of N_2 can be seen in Figure B. 4. The species N_2^+ would have one less bonding electron and N_2^- would have an additional electron located in a π_{2p}^* molecular orbital. The net bond order of N_2 is 3 and of N_2^+ and N_2^- is each 2.5. Both N_2^+ and N_2^- are less stable than N_2 and both would have a longer bond length than N_2.

B.7 See Figure B.4 . The O_2 molecule has the net bonding of a double bond while the O_2^{2-} ion has the net bonding of a single sigma bond. Therefore, the bond order of O_2 is 2 and that of O_2^{2-} is 1. The bond energy of O_2 would be greater, its bond length would be less, and its vibrational frequency would be greater than that of O_2^{2-}.

B.9 (a) Because a bonding electron is removed in each case, Li_2^+, B_2^+, and C_2^+ would be less stable than the neutral X_2 species. Be_2^+ would be more stable than Be_2 because an antibonding electron is removed.
(b) Li_2^- would be less stable, Be_2^-, B_2^-, and C_2^- would be more stable than the neutral X_2 species. The extra electron is antibonding in Li_2^- but is bonding in Be_2^-, B_2^-, and C_2^-.

B.11 H_2^+ Bond order = 1/2; $(\sigma_{1s})^1$
He_2^+ Bond order = 1/2; $(\sigma_{1s})^2 (\sigma_{1s}^*)^1$